Fish on Paper

Emergence of Natural History

VOLUME 10

The titles published in this series are listed at *brill.com/enh*

Fish on Paper

Ichthyology and the Disciplining of Natural History (1680–1820)

By

Didi van Trijp

BRILL

LEIDEN | BOSTON

This research has been funded by the Netherlands Organisation for Scientific Research (NWO, project number 360-55-090). Open Access of this publication was made possible by additional funding of NWO.

Chapter 2 was published as Didi van Trijp, "Fresh Fish: Observation up Close in Late Seventeenth-Century England." *Notes and Records of the Royal Society of London* 75, no. 3 (2021): 311–332.

Cover illustration: Flying fish (*Exocoetus evolans*). In Marcus Elieser Bloch, *Ichtyologie, ou, Histoire naturelle, générale et particulière des poissons : avec des figures enluminées, dessinées d'après nature* (Berlin, Bloch – De la Garde: 1785–1797 [1797]), vol. 12, pl. 398. The New York Public Library, Rare Book Division.

The Library of Congress Cataloging-in-Publication Data is available online at https://catalog.loc.gov
LC record available at https://lccn.loc.gov/2025038914

Typeface for the Latin, Greek, and Cyrillic scripts: "Brill". See and download: brill.com/brill-typeface.

ISSN 2452-3283
ISBN 978-90-04-54949-4 (hardback)
ISBN 978-90-04-54950-0 (e-book)
DOI 10.1163/9789004549500

This book is printed on acid-free paper and produced in a sustainable manner.

Contents

Acknowledgements

When starting this project, I could not have fathomed the many brilliant people I would meet, the unique historical materials I would marvel at, or the fascinating exchanges it would spark. I am grateful for all of it. I want to thank some of the people to whom I am particularly indebted here.

This book arises from my doctoral research, carried out at Leiden University as part of the project 'A New History of Fishes. A Long-Term Approach to Fishes in Science and Culture 1550–1880', funded by the Dutch Research Council (NWO). First and foremost, my gratitude goes out to my supervisors, Paul J. Smith, Eric Jorink and Nadine Akkerman for their invaluable support and advice over the years, as well as for their sharp attention to argument and delightful detail. It has made me a better historian.

The Leiden University Centre for the Arts in the Society (LUCAS) has provided a stimulating, cross-disciplinary setting for this research, offering everything that a doctoral student might need. I thank my fellow project members, Florike Egmond, Marlise Rijks, Sophia Hendrikx and Robbert Striekwold for many inspiring conversations. Thanks also to my other colleagues in Leiden, most of all Lieke Smits, Bareez Majid, Anna Volkmar and Lisanne Wepler. In addition, I want to express thanks to my colleagues at the editorial offices of *Isis* at Utrecht University where I worked during part of this project; especially to Floris Cohen, Desiree Capel and Ad Maas, for creating a lively group that I am still glad to be a part of.

Several institutions have made this research possible. I am grateful to the British Society for the History of Science for funding archival research in the Nottingham University Library, to the Royal Society for facilitating research in various libraries and archives in London through the Lisa Jardine Research Award, and to the Francke Foundations for enabling perusal of their archives through the Dr. Liselotte Kirchner Fellowship. My heartfelt thanks to the staff of all the libraries, archives and museums that I have consulted in the past years, without whom this research would not at all have been possible. In particular, I want to thank Virginia Mills for introducing me to the Royal Society archives, Isabelle Charmantier for welcoming me at the Linnean Society, James Maclaine for accommodating my research in the Natural History Museum, and Edda Aßel for receiving me at the Museum für Naturkunde.

Audiences at conferences in Keulen, Halle, Augsburg, Berlin, Oxford, Paris, Amsterdam, Utrecht, Woudschoten and Rolduc have helped me to sharpen my ideas. I have had the privilege, furthermore, to attend not one but two summer schools at the University of Göttingen, organised by Dominik Hünniger,

Marie Luisa Allemeyer and Christian Vogel at the congenial setting of the Lichtenberg-Kolleg. The insightful feedback of my commentators, Emma Spary and Joachim Rees, as well as the general discussions among the participants have been instrumental in shaping my views on material culture and the history of collecting.

For engaging conversations on early modern science and natural history, I am grateful to Katherine Reinhart, Felicity Henderson, Sietske Fransen, Sachiko Kusukawa, Philippa Hellawell, Meike Knittel, Simon Rebohm, Karsten Hommel, Stéphane Schmitt, Hans Aili, Anita Guerrini, Jan Overwijk, Azadeh Achbari, Ruben Verwaal, Jessie Wei-Hsuan Chen, Eulàlia Gassó Miracle, Esther van Gelder, Holger Zaunstöck, and Markus Berger. Martien van Oijen kindly answered any fish-related questions I had. Marc Argeloo and Johannes Müller generously shared their knowledge on the critical nexus of history, ecology and the environment. Mathijs Boom, Marieke Hendriksen and Wijnand Mijnhardt offered incisive comments on versions of chapters. I am especially indebted to Robbert Striekwold, whose name deserves mentioning twice in these acknowledgements for reading several drafts of what would become this manuscript with an ever critical yet constructive eye. Many thanks are due to Hilke Hoogenboom for diligent translations from Latin, and to Pete Langman for meticulous editing.

I am grateful to the series editors of *Emergence of Natural History* who helped me to envision this research as a book, first Kay Etheridge and later Andreas Weber, and who offered useful advice at crucial stages. My thanks also to the two anonymous reviewers, whose thoughtful comments have significantly strengthened this work. At Brill, Alessandra Giliberto, Melissa Allieri and Gera van Bedaf guided me expertly through the publication process.

Friends and family made sure I stayed afloat. For this, I thank Lisanne, Sophie, Frédérique, Sarah, and Nina. I am grateful to my parents, Ko and Anja, for unconditional support. And, finally, for his encouragement and love, and for never failing to make me laugh even at moments when such a thing did not seem likely, I thank Sven.

Figures

Abbreviations

Archives, Libraries and Collections

AFSt	Archiv der Franckeschen Stiftungen (Halle an der Saale)
BBAW	Berlin-Brandenburgische Akademie der Wissenschaften (Berlin)
PAW	Preußische Akademie der Wissenschaften
BL	British Library (London)
Add	Additional
MS	Manuscript
KB	Koninklijke Bibliotheek (The Hague)
KBS	Kungliga Biblioteket (Stockholm)
LS	Linnean Society (London)
MfN	Museum für Naturkunde (Berlin)
GNF	Gesellschaft Naturforschender Freunde
ZMB	Bestand Zoologisches Museum
NHM	Natural History Museum (London)
NUL	Nottingham University Library (Nottingham)
RS	Royal Society (London)
SA	Stadsarchief Amsterdam (Amsterdam)
SBB	Staatsbibliothek zu Berlin (Berlin)
SUB	Stockholms Universitetsbibliotek (Stockholm)
UBA	Universiteitsbibliotheek Amsterdam (Amsterdam)
UBL	Universiteitsbibliotheek Leiden (Leiden)
UL	Universitätsbibliothek Leipzig (Leipzig)
UUB	Uppsala Universitetsbibliotek (Uppsala)

Other

Allg. Nat. der Fische	Marcus Élieser Bloch, *Allgemeine Naturgeschichte der Fische* (Berlin: Hesse, Realschule and J. Morino, 1782–1795)
aph.	aphorism
Hist. anim. IIII	Conrad Gessner, *Historiae animalium liber* IIII (Zurich: Christoph Froschauer, 1558)
Hist. pisc.	Francis Willughby and John Ray, *Historia piscium* (Oxford: Sheldonian Theatre, 1686)
Icht.	Peter Artedi, *Ichthyologia, sive opera omnia de piscibus* (Leiden: Conrad Wishoff, 1738)
OFB	Oxford Francis Bacon

Notes

On Translations

Most of the translations have been made by the author. If existing translations have been adapted or expanded, this is indicated in the footnotes. The Latin translations have been checked and corrected by Hilke Hoogenboom. In most cases, the '&' has been replaced with 'and' in order to improve readability.

On Dates

Dates given conform to the Gregorian calendar and thus are New Style (NS), except when marked OS (Old Style) to accommodate the Julian calendar that remained in use in England until 1752. When giving dates in OS, the English traditional new year of 25 March (Lady Day) rather than 1 January is observed.

On Transcriptions

In passages transcribed from letters and manuscripts, the spelling, punctuation, capitalisation and italics have been rendered faithfully. The long 's', however, has been normalised for readability. If emphasis has been altered or added, this is indicated in the footnote.

On References

In citing unpublished source material, reference is made to folio 'f', with indication of the recto 'r' or verso 'v' side of the page.

In citing primary printed works, the page numbers are given where possible. Where no page number is available, as is often the case with prefatory material, for example, references are given by signature (sig.). In the case that no pagination is offered and signature references are also lacking, this is signalled with n.p.

On Species Names

In those cases where the currently accepted scientific name for a species can be ascertained, this name is mentioned in the footnote.

Introduction: No Such Thing as a Fish

There is no such thing as a fish. A recent encyclopaedia of aquatic life has raised the question of whether we can know a fish when we see one, submitting that "the concept is merely a convenient umbrella term to describe an aquatic vertebrate that is not a mammal, a turtle, or anything else."[1] This makes sense from the perspective of phylogenies, i.e. the evolutionary histories of and relationships between groups of species. For if we look closely at bony fish like the tuna, lobe-finned fish like the coelacanth, cartilaginous fish like the shark, or jawless fish like the lamprey – really closely, by examining their DNA, for example – we see that they are in fact far too diverse to be grouped all together under the banner 'fish'.[2] As it turns out, they are often more closely related to other vertebrates than to each other. To put it bluntly, tuna are more closely related to humans than they are to sharks.[3]

And yet, that does not mean that the term fish does not carry any meaning. It has been in use for centuries to denote the diverse denizens of oceans, seas, rivers, streams, ponds and lakes, and is still widely used today. In fact, the question of what, exactly, comprises a fish, and how best to approach this animal as a subject of research, itself is a historical question. The long eighteenth century, a period preoccupied (if not obsessed) with creating order in nature, saw a particularly rich succession of different definitions for these animals. At stake in this question of defining what precisely a fish was, were matters like the value that should be attached to observation and experience, and what precisely these concepts constituted. What methods were best suited to study the slippery subject of fish, and who, precisely, could be counted on to produce solid, authoritative knowledge about it? Over the course of the 'long' eighteenth century (taken here as the period from the 1680s to the 1820s), several scholars occupied themselves with carefully describing all fish hitherto known and painstakingly attempted to place them into their proper order. The 'fish books' that they subsequently published offered up classifications, descriptions and, at times, stunning illustrations.

1 *The Encyclopedia of Underwater Life*, comp. Andrew Campbell and John Dawes (Oxford: Oxford University Press, 2005), s.v. "Fish, What is A?".

2 For the impact of genetic analyses on taxonomic divisions, see: Thomas J. Near and Christine E. Thacker, "Phylogenetic Classification of Living and Fossil Ray-Finned Fishes (Actinopterygii)," *Bulletin of the Peabody Museum of Natural History* 65, no. 1 (2024): 340–357.

3 Jonathan Balcombe, *What A Fish Knows* (London: Oneworld, 2019), 12.

This book marks out the long eighteenth century as a period in which ichthyology was shaped as a result of the 'disciplining' of natural history. It argues that this disciplining took place primarily through the practice of classifying. First of all, the ichthyologist was, through the classificatory method, elevated as a specific, methodic authority on fish, and set apart from practical, artisanal communities of fishermen, fishmongers and cooks who had privileged access to the world underwater as a site of investigation. Secondly, establishing what 'formal fish knowledge' entailed also served to envision it as a separate branch of natural history. The authors discussed in this work sought to present themselves as authoritative knowers of fish through the formulation of specific approaches to collecting, describing, illustrating and classifying specimens – assessing, for example, the value of examining the preserved skin of a fish against a coloured illustration of it. Such debates both shaped and were shaped by this process of disciplination. In paying attention to these different acts of demarcation, this book offers the first contextualised approach to how ichthyology took shape as a distinct field of knowledge in this period.

1 Disciplinary Histories

The leaky category that is 'fish', however, should give us pause to think and question our preconceived ideas about traditional histories of disciplines. The eighteenth century has been described as a time of fermentation, in which both ideas about how nature should be studied as well as how the study of nature should be organised underwent significant changes.[4] An impressive range and volume of books were published ranging from studies of plants and animals and new ideas about medicine.[5] At the same time, we are reminded by Richard Yeo that "people in the eighteenth century did not share our modern sense of the scope and boundaries of scientific subjects. They certainly did not recognise the closely differentiated array of disciplines, often marked by special journals and institutions, that began to emerge in the early nineteenth century."[6] Thus, while on the one hand the study of living nature intensified and naturalists came to apply themselves to specific topics within this field,

4 See: George Sebastian Rousseau and Roy Porter, eds., *The Ferment of Knowledge: Studies in the Historiography of Eighteenth-Century Science* (Cambridge: Cambridge University Press, 1980).

5 Anita Guerrini, "The Material Turn in the History of Life Science: Life Sciences," *Literature Compass* 13, no. 7 (2016): 470.

6 Richard Yeo, "Classifying the Sciences," in Porter, *Cambridge History of Science*, vol. 4, 243–244.

on the other, these investigations did not crystallise into the clearly separated disciplinary categories that we know today. It is on this murky process that this book sheds light.

The concept of 'discipline' has its etymological roots in the Latin *disciplinas*, which is derived from the verb *discere* (to learn). As Donald R. Kelley has explained, its early modern usage mostly stemmed from the context of pedagogy.[7] Today, the word 'discipline' has come to denote an autonomous field of knowledge. Although the question of how modern disciplines came into being has long held the interest of historians of science, there is no uniform understanding of what our contemporary concept of 'discipline' encompasses, and neither has it been consistently applied.[8] Historians generally situate the formation of distinct disciplines in the nineteenth century. The markers they developed for considering a field of knowledge an actual discipline relate both to the content of a discipline (for example, the formulation of shared questions, methods and goals) and to its organisational structure (for example, designated spaces and institutional embedding in the form of appropriate roles, journals, courses of study, textbooks, and dedicated societies).[9]

If the nineteenth century was the age of discipline formation, how do we interpret what happened during the centuries prior? Here, too, fields of knowledge emerge – or rather, are given shape – from the study of nature as a whole. Some historians see this as the time when the founders of disciplines were born: bright minds, well ahead of their time, who paved the way to the modern science we know today. While it is all too tempting to transpose our current ideas of a specific discipline onto that of the past, it is important to examine the formation of disciplines without taking their current form or their supposed coherence for granted.[10] Here, the notion of 'boundary work' – a notion that sociologists of science use for the active process in which scientists differentiate what they consider science from non-science – can be a useful place to

7 Donald R. Kelley, "The Problem of Knowledge and the Concept of Discipline," in *History and the Disciplines: The Reclassification of Knowledge in Early Modern Europe*, ed. Donald R. Kelley (New York: University of Rochester Press, 1997), 15.

8 Rudolf Stichweh, "The Sociology of Scientific Disciplines: On the Genesis and Stability of the Disciplinary Structure of Modern Science," *Science in Context* 5, no. 1 (1992): 7; Kelley, "Introduction," in *History and the Disciplines*, 1.

9 Ibid., 9.

10 Azadeh Achbari, "Rulers of the Winds: How Academics Came to Dominate the Science of Weather" (PhD diss., Vrije Universiteit Amsterdam, 2017), 227.

start.[11] Such a view of disciplines as carefully constructed entities helps to open up research on the process of specialisation.

In the history of science, two books explicitly tackle the issue of discipline formation within natural history. These are Roy Porter's *The Making of Geology: Earth Science in Britain, 1660–1815* and Paul Lawrence Farber's *The Emergence of Ornithology as a Scientific Discipline, 1760–1850*. Both seek out to understand how the study of the earth and of birds respectively coalesced into recognisable disciplines. Farber's main aim is to understand how ornithology became a discipline "that attracted a substantial group of scientists who shared a set of rigorous methods, exacting criteria, and ambitious goals."[12] In answering this question, his primary focus is on debates around classification. What precisely counted as a rigorous method, however, in terms of actual practice remains out of sight. In contrast, Porter is concerned with showing how practitioners of the emerging field of 'geology' wanted it to become more and more 'scientific' while the very criteria for what counted as 'scientific' were themselves also under review.[13] In other words, enthusiasts for the study of the earth set up rules for how to do 'geology' while simultaneously changing these rules based on their continuous discussion of the matter.

Porter examines how the study of the earth was made into what in the early nineteenth century came to be called geology. In order for this to happen, the earth had to be moulded into an object of scientific study that behaved according to natural laws which could be uncovered by examining its crust. Shifts in terminology, he claims, require attention as "the coining of the term 'geology' and its derivatives" indicate "changing ways of seeing."[14] The present book, likewise, interrogates the term ichthyology as used by historical actors. During the late seventeenth century, 'ichthyology' was a far from current term. This is why, even though Conrad Gessner (1516–1565) published a lavish volume on the natural history of creatures living in and near the water, that does not mean that he was an ichthyologist, or that we should refer to his work as an ichthyological one. The term did gain currency over the course of the

11 The notion of "boundary work" is defined as "an ideological style found in scientists' attempts to create a public image for science by contrasting it favorably to non-scientific intellectual or technical activities" in Thomas F. Gieryn, "Boundary-Work and the Demarcation of Science from Non-Science: Strains and Interests in Professional Ideologies of Scientists," *American Sociological Review* 48, no. 6 (1983): 781.

12 Paul Lawrence Farber, *The Emergence of Ornithology as a Scientific Discipline, 1760–1850* (Dordrecht: Springer, 1982), xxi.

13 Roy Porter, *The Making of Geology: Earth Science in Britain, 1660–1815* (Cambridge: Cambridge University Press, 1977), 5.

14 Ibid., 4.

eighteenth century as a way of denoting a particular way of examining fishes. This book therefore examines the term 'ichthyology' with an eye to the way in which historical actors themselves defined it, why they wielded it and how it gained currency as a marker of a certain kind of knowledge about fish. As we will see in Chapter 3, the Swedish naturalist Peter Artedi did this both deliberately and exhaustively.

Remarkably, neither of the two monographs includes any of the many images that were produced to facilitate the study of birds as well as fossils, minerals and stones over the course of the long eighteenth century. Farber's monograph on ornithology only includes portraits of eminent naturalists, and Porter's careful, constructionist account of the study of geology does not incorporate any visual material at all. This is admittedly strange for two domains of study which, in the early modern period, relied so heavily on the use, circulation and study of images. That this seems such an oversight, however, is indicative of the rapid rise of the interest in visual culture within the history of science over the last two decades.[15] Images are now recognised as important tools of research that serve argumentative, descriptive and demonstrative functions.[16] In fact, the practice of illustration (alongside other practices of natural history, such as collection, description and classification) lay at the core of early modern natural history. The reasons why it is worth focusing on the epistemological discussions that surrounded these practices will be explored in detail later in this introduction.

Such epistemological discussions always took place in a social context. As Emma Spary has argued, claims of natural historical knowledge were negotiations in which it was not only the identification of a particular species that was at stake, but also the authority and credibility of the person who described, depicted and/or classified it. Settling the identity of a specimen or even determining the value of an entire classification system therefore meant assessing the value of the naturalist who proposed it.[17] The following section looks more closely at how social status related to authority. In line with recent research that has brought to light the indispensable role played by people who have

15 For an overview of this development, see: Alexander Marr "Knowing Images," *Renaissance Quarterly* 69, no. 3 (2016): 1000–1013.

16 Susan Dackerman, "Introduction: Prints as Instruments," in *Prints and the Pursuit of Knowledge*, ed. Susan Dackerman (Cambridge, Mass.: Harvard Art Museums, 2011), 28.

17 Emma Spary, "Of Nutmegs and Botanists: The Colonial Cultivation of Botanical Identity," in *Colonial Botany: Science, Commerce, and Politics in the Early Modern World*, eds. Londa Schiebinger and Claudia Swan (Philadelphia: University of Pennsylvania Press, 2005), 187–203.

traditionally fallen outside the purview of studies of natural history, it argues for considering a wider cast of naturalists.

2 Angling for Authority

This book examines how naturalists asserted themselves as authoritative knowers of fish: individuals who could be trusted upon to produce reliable knowledge about these slippery creatures. Works by Steven Shapin, Simon Schaffer and many others since have firmly established the notion that the credibility of a knowledge claim ultimately rested on the credibility of the person making it. The issue of credibility was never far away when it came to the making of early modern natural knowledge. Shapin showed this in highlighting how Robert Boyle (1627–1691) engaged with divers as he was trying to understand the pressure exerted by water. An accomplished, respectable member of the Royal Society, Boyle admitted his own experience with water was surface level: "I do not pretend to have visited the bottom of the sea; [and] none of the naturalists whose writings I have yet met with, have been there any more than I."[18] Where Boyle initially saw divers as the most reliable witnesses because of their sustained experience underwater, he came to characterise them as unreliable, vulgar observers when their statements repeatedly contradicted his hypothesis about the gravitational properties of water.

We should, however, not regard the interactions between Boyle and divers as merely revolving around social standing. As Phillippa Hellawell argues, in his wider work, Boyle tended to rank testimonies based on the knowledgeability and skill of those who provided it, not simply rejecting such accounts because he deemed them to originate from low-ranking informants, but rather because he thought they were made by "persons void of curiosity and skill to make such observations".[19] Besides social standing, then, other significant factors were curiosity, skill, occupation and experience. This book will examine how multiple markers of credibility were simultaneously in play when it came to the study of the world underwater.

For aspiring naturalists, the matter of authority was especially pressing because building a career around the study of nature was far from a straightforward matter in the early modern period. Early modern career trajectories

---

18 As cited in Steven Shapin, *A Social History of Truth: Civility and Science in Seventeenth-Century England* (Chicago: Chicago University Press, 1994), 260.

19 Philippa Hellawell, "'The Best and Most Practical Philosophers': Seamen and the Authority of Experience in Early Modern Science," *History of Science* 58, no. 1 (2019): 32.

can seem to our eyes both "dazzlingly diverse and oddly circumscribed", in the words of Katharine Park and Lorraine Daston.[20] We might assume that research was primarily conducted at universities, but these were more focused on teaching; if professors wished to do research, they had to do so in their spare time. Research was also carried out in what we would now consider less expected places. Artisans and apothecaries examined and experimented with the properties of all kinds of materials in workshops and kitchens. Mathematicians and physicians, furthermore, found occasional patronage in courts or estates with observatories, menageries and lavish gardens. Those early moderns who were bent on dedicating their life to the study of living nature had different routes available to them, but could seldom take one effortless, obvious path towards comfortable remuneration.

In the absence of what we might consider clearly recognisable jobs, let alone job titles, how ought we best refer to that diverse crew of early moderns busying themselves with the study of nature? Historians have adopted the term 'naturalist' as an apt term to capture an individual who busied him or herself with the study of nature in the early modern period. Historians of science have taken to using this designation as it is relatively close to the categories that were used by historical actors themselves.[21] The term is sufficiently broad to apply to those early moderns who studied the general properties of nature as well as to those investigating its particularities – and of course to those who did both. The study of nature was far from the exclusive domain of professors or physicians, but also taken up by apothecaries, women healers, herbalists, gardeners, draughtsmen and goldsmiths, and so forth. The "concerted program of boundary expansion"[22] that Deborah Harkness has signalled in her book on the production of knowledge in Elizabethan London, continues still. Historians are more explicitly incorporating the contributions and perspectives of those who have fallen outside of the traditional purview of the history of science into the narrative.[23] It has required them to let go of preconceived ideas of what does

20 Katharine Park and Lorraine Daston, "Introduction," in *Cambridge History of Science*, vol. 3: *Early Modern Science*, eds. Lorraine Daston and Katherine Park (Cambridge: Cambridge University Press, 2006), 5.

21 Such as the term 'studiosus rerum naturae' used for those who were devoted to the study of natural matters; see: Park and Daston, "Introduction," 10; Brian Ogilvie, *The Science of Describing: Natural History in Renaissance Europe* (Chicago: University of Chicago Press, 2006), 54.

22 Deborah E. Harkness, *The Jewel House: Elizabethan London and the Scientific Revolution* (New Haven: Yale University Press, 2007), 255.

23 See, for example: Anna Winterbottom, *Hybrid Knowledge in the Early East India Company World* (Basingstoke: Palgrave Macmillan, 2015); Jaime Marroquin Arredondo and

and what does not constitute natural knowledge, and has often required some industry on their part to retrieve past efforts that are no longer readily visible. As Pamela Smith has demonstrated, artisanal knowledge, which was heavily reliant on bodily experience, came to be absorbed by the work of the natural philosopher, while the artisans themselves were gradually expelled from it.[24]

A broad variety of natural historical practitioners also inhabit the sources that form the corpus of this study. They are unnamed Tamil divers, tasked to retrieve rare shells from the Bay of Bengal, or pearls and other collectables from the deep; they are artisans working with fish, like the Scheveningen fishmonger Adriaen Coenen (1514–1587) or the Strasbourg fisherman Leonhard Baldner (1612–1694). These men, whose initial experience with fish derived from hook and line rather than from the written word, produced stunning illustrated manuscripts describing the water creatures they came across. In these manuscripts they shared their observations and put forward hypotheses regarding complicated questions. Baldner, for example, positioned himself in opposition to the esteemed naturalist Conrad Gessner in the debate surrounding the procreation of carp by saying that they did not generate spontaneously from mud, but all came from roe. Or they are Indian drawing masters on the coast of Coromandel who produced rich illustrations of fishes but were rarely named naturalists in their own right.

Often overlooked are the observations women made of the behaviour of species, as well as how they collected and circulated specimens. Some such as the French printmaker Marguerite Lecomte (1717–1800) who engraved plates for one of the most elaborate taxidermy manuals of the eighteenth century, Turgot's *Mémoire instructif* (Figure 1) discussed below, played a vital role in the production of books. More anonymous, but equally important, were the women employed throughout Europe as colourists for engraved plates. Women of means were important patrons in the publication of natural historical books. Their means to assert authority in the examination of nature, however, were more limited than those of men as they were invariably not granted access to places where knowledge of nature was formally taught or discussed, such

Ralph Bauer, eds., *Translating Nature: Cross-Cultural Histories of Early Modern Science* (Philadelphia: University of Pennsylvania Press, 2019); Anna Marie Roos, *Martin Lister and His Remarkable Daughters: The Art of Science in the Seventeenth Century* (Oxford: The Bodleian Library, 2019); Arlene Leis and Kacie L. Wills, eds., *Women and the Art and Science of Collecting in Eighteenth-Century Europe* (New York: Routledge, 2020).

24 Pamela Smith, *The Body of the Artisan: Art and Experience in the Scientific Revolution* (Chicago: University of Chicago Press, 2004), 186.

as universities or learned societies.[25] Their study of nature often took place within a domestic context, where various members of the household tended to the examination of nature as a collaborative project.[26]

Accommodating as the term 'naturalist' may be for capturing the variety of people that examined nature, the way it is used can still be discriminatory. Those naturalists on whom historians fix their gaze have ever been, and largely remain, those men of higher social standing who possessed sufficient private means to carry out their studies, and to publish books about it. In their reports, these gentlemen often refer to the suppliers of their *naturalia* and/or those experts they consulted in a somewhat offhand manner. As Lydia Barnett has argued, eighteenth-century gentlemen naturalists credited the fossil finders on whom they depended in a selective manner, often through anonymous acknowledgement.[27] This also caused tension, because who, then, had actually discovered a particular fossil? Naturalists eased such anxieties by qualifying these people of practice as ignorant and incompetent, thus allowing them to assert themselves as the sole discoverers and interpreters of natural facts. Although it sometimes requires reading against the grain, this book seeks to draw out the rich and vibrant culture of examination in which early modern natural history was embedded.

Bound up in the epistemological world of fish is an entire social world that we miss when we are too preoccupied with tracing the development of disciplines. Focusing on practices is a good way to draw out this shared world, and by closely examining the debates underpinning them, we get a better picture of how early modern naturalists angled for authority. When we centre practices, we create space for a wider cast of historical actors having engaged in the study of natural history by collecting, observing, describing, illustrating and classifying.

3 Practices of Natural History

To disregard practices such as the illustrating of specimens is to omit debates that lay at the core of natural historical study in the early modern period.

25 Londa Schiebinger, "Women of Natural Knowledge," in Daston and Park, *Cambridge History of Science*, vol. 3, 195.

26 Elaine Leong, *Recipes and Everyday Knowledge: Medicine, Science, and the Household in Early Modern England* (Chicago: Chicago University Press, 2018), esp. 46–70.

27 Lydia Barnett, "Showing and Hiding: The Flickering Visibility of Earth Workers in the Archives of Earth Science," in *History of Science* 58, no. 3 (2020): 254.

Images often functioned as a means of communication. At times, they even served as evidence. Both creating images and evaluating their accuracy included many steps, however minor they might seem from a distance, which directly corresponded to fundamental debates around how naturalists could trust to come to any sound conclusion on their object of study based on their interpretation of these materials. In the case of fish, for example, one stroke of the pencil – thus giving a fish a fin ray more, or less – could have repercussions for how the depicted species would be classified. Early moderns were acutely aware of both the importance and epistemological instability of images and discussed such issues at length. This book is concerned with precisely such discussions. It concentrates on four important practices: those of collection, description, illustration and classification. The benefit of focusing on practices is that it helps to reframe the production of natural knowledge as a process rather than as an outcome.[28]

The first practice to discuss is that of **collection.** Examination of nature, after all, often started with going into the field – or indeed onto or even into the water – to observe as well as to collect. As the French naturalist René-Antoine Ferchault de Réaumur stated, "persons who have at heart the progress of natural history, and intended to facilitate the study of it, must needs be desirous to see the collections of diverse sorts of productions."[29] The eighteenth century in particular saw a surge in instructions for the preparation of specimens. The aforementioned *Mémoire instructif* by Étienne-François Turgot (1721–1789) is unique in its minute attention to the steps entailed in turning species into specimens.[30] The plates depict scalpels and other instruments cleaving through bodies, or the incision marks left in bodies that have been stitched back up again (Figure 1). It is manuals like this one that allow historians to examine taxidermy strategies and even re-enact them.[31] Such hands-on efforts offer insight into how specimens themselves, like their descriptions or

28 James Secord, "Knowledge in Transit," *Isis* 95, no. 4 (2004): 658.

29 René-Antoine Ferchault de Réaumur, "Divers Means for Preserving from Corruption Dead Birds, Intended to Be Sent to Remote Countries, So That They May Arrive There in a Good Condition. Some of the Same Means May be Employed for Preserving Quadrupeds, Reptiles, Fishes, and Insects, by M. de Reaumur, F.R.S. and Memb. Royal. Acad. Sc. Paris. Translated from the French by Phil. Hen. Zollman, Esq; F.R.S." in *Phil. Trans.* vol. 45 (1748): 305.

30 Étienne-François Turgot, *Mémoire instructif sur la manière de rassembler, de preparer, de conserver, et d'envoyer les diverses curiosités d'histoire naturelle* (Lyon: chez Jean Marie Bruyset, 1758).

31 Marieke M.A. Hendriksen, "Animal Bodies between Wonder and Natural History: Taxidermy in the Cabinet and Menagerie of Stadholder Willem V (1748–1806)," *Journal of Social History* 52, no. 4 (2019): 1110–1111.

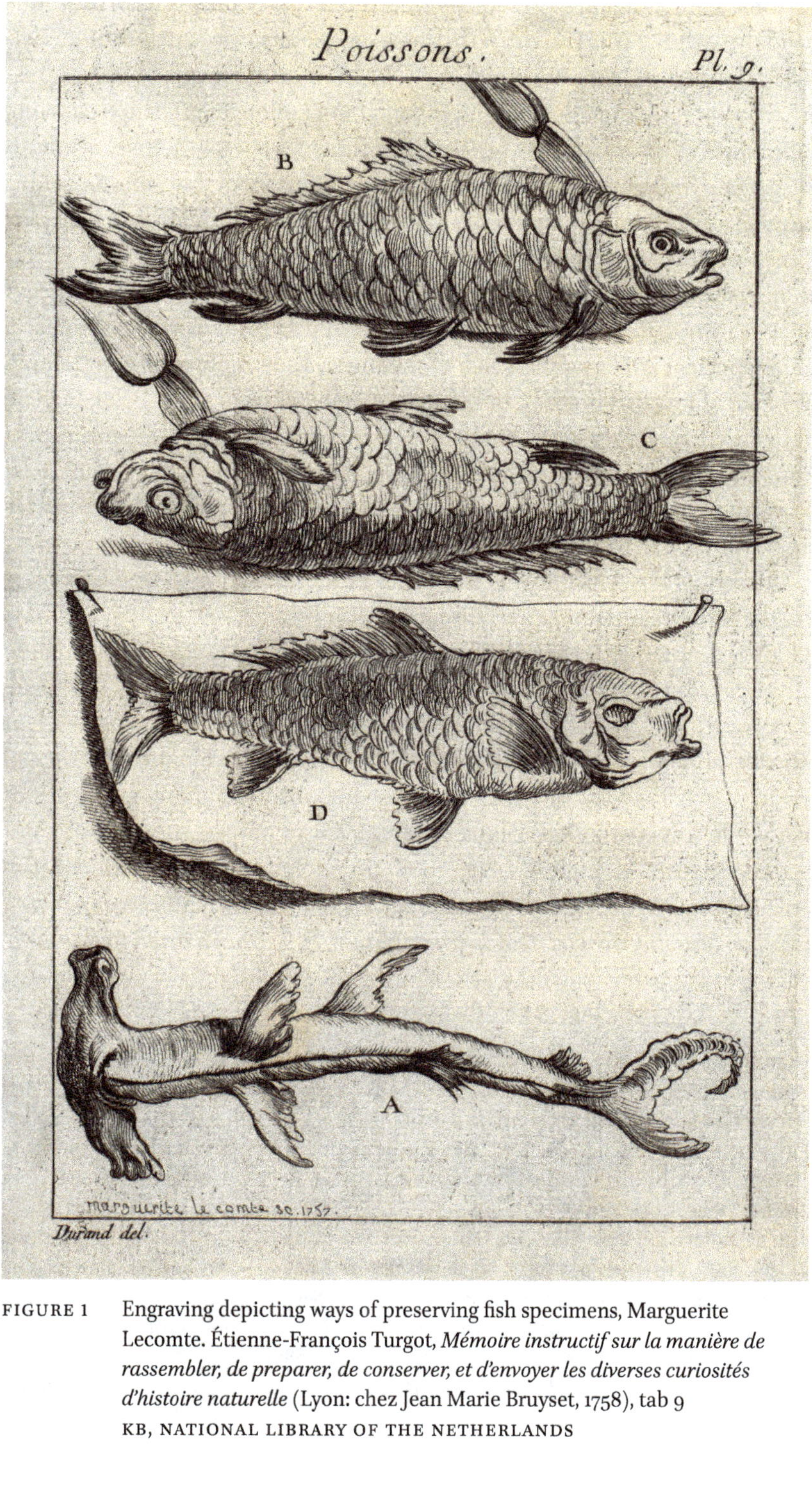

FIGURE 1 Engraving depicting ways of preserving fish specimens, Marguerite Lecomte. Étienne-François Turgot, *Mémoire instructif sur la manière de rassembler, de preparer, de conserver, et d'envoyer les diverses curiosités d'histoire naturelle* (Lyon: chez Jean Marie Bruyset, 1758), tab 9
KB, NATIONAL LIBRARY OF THE NETHERLANDS

illustrations, were (and are still) selectively edited. This realisation corresponds more widely with the material turn that envisages objects as carriers of knowledge with their own uses and limits.[32] Apart from creating specimens, transporting them across long distances was a further challenge: the ships carrying them endangered by storms or pirates, and the specimens themselves damaged by vermin and neglected by crews.[33] No wonder, then, that self-conscious collectors used cabinets, rooms and drawers filled with specimens to display their possession of, and command over, nature.[34]

The next practice is that of **description**. Both the content and shape of description were guided by the question: what did one see when looking at a particular species? An entangled web of meaning, a set of physical characteristics, processes like reproduction or behaviour, or something else? Brian Ogilvie has argued that the emergence of natural history in the sixteenth century was born out of a preoccupation with description, a practice that was systematised in the seventeenth century.[35] For many naturalists who concerned themselves with the study of fish, physical attributes were of primary interest. Even if natural historical description was moulded into conventions, it nevertheless remained a topic of reflection and discussion.

A following practice is that of **illustration**, which due to its significance to natural history has been touched upon a few times earlier in this introduction. That this practice was closely related to that of collecting becomes clear from the example of an engraving that the physician Hans Sloane (1660–1753) had made of an albacore, or white tuna. Having observed many examples of this fish while traversing the oceans to Jamaica as a part of the entourage of the second Duke of Albemarle (1653–1688), he included an engraving of it in his book *A voyage to the islands Madera, Barbados, Nieves, S. Christophers and Jamaica*. The illustration was made, he explained, "[...] from a dried Fish, where every thing was perfect save the first Fin on the Back, which I suppose was

32 Sven Dupré and Christoph Lüthy, "Introduction," in *Silent Messengers: The Circulation of Material Objects of Knowledge in the Early Modern Low Countries*, eds. Sven Dupré and Christoph Lüthy (Berlin: LIT, 2011), 12.

33 Christopher M. Parsons and Kathleen S. Murphy, "Ecosystems under Sail: Specimen Transport in the Eighteenth-Century French and British Atlantics," *Early American Studies* 10, no. 3 (2012): 538.

34 Paula Findlen and Anna Toledano, "The Materials of Natural History," in *Worlds of Natural History*, eds. Helen Anne Curry, Nicholas Jardine, James Secord and Emma C. Spary (Cambridge: Cambridge University Press, 2018), 151. See also: Oliver Impley and Arthur MacGregor, eds. *The Origins of Museums: The Cabinet of Curiosities in Sixteenth- and Seventeenth-Century Europe*. Oxford: Clarendon Press, 1985.

35 Brian Ogilvie, *The Science of Describing: Natural History in Renaissance Europe* (Chicago: University of Chicago Press, 2006), 7.

accidentally rub'd off."[36] The illustration restored this fin, but its absence might easily have slipped past a less attentive observer. In line with recent research on the visual culture of early modern science, this book analyses the functions and uses of images, by regarding these as epistemological tools that are shaped and driven by practical, pragmatic and financial incentives.[37]

The last practice discussed here is that of **classification.** It is perhaps because classification has since come to occupy such a central place in natural history that internalist narratives have been produced around it. These take the form of assessments of whether or not people of past times were the first to describe a certain species, and whether they had assigned it to its 'rightful' place in the larger system of classification (according to our contemporary standards). Over the last decades, however, historians have come to consider precisely what it was that naturalists sought to accomplish by classifying nature, as well as how they went about it. This means that these historians no longer perceive classification necessarily as a neutral act, but as one that reflects certain cultural and social values.[38] Furthermore, over the course of the eighteenth century the act of classifying nature itself became considered paramount for what it meant to be a naturalist.

An additional advantage to the study of practices is that it elucidates how plants and animals did not passively submit themselves to natural historical inquiry. For example, plants might produce toxic sap once picked, whereas certain species of fish, like pufferfish, have venomous spines that paralyse those who touch them. Animals resisted capture and transport, and sometimes even inflicted damage after death by withstanding straightforward preparation. As such, their way of being – the very thing that naturalists sought to understand – in fact problematised the methods employed to understand it by naturalists.[39] In her book *Curious Species*, Whitney Barlow Robles has shown the many ways in which (nonhuman) animals acted upon the world, arguing that "[…] fish and other creatures circumscribed what could be visualized, stabilized, and

36 Hans Sloane, *A voyage to the islands Madera, Barbados, Nieves, S. Christophers and Jamaica with the Natural History of the Herbs, and Trees, Four-footed Beasts, Fishes, Birds, Insects, Reptiles &c. of the Last of Those Islands*, vol. 1 (London: B.M. for the Author, 1707), 11.

37 Sachiko Kusukawa, "Illustrating Nature," in *Books and the Sciences in History*, eds. Nicholas Jardine and Marina Frasca-Spada (Cambridge: Cambridge University Press, 2000), 109.

38 See: Londa Schiebinger, "Why Mammals are Called Mammals: Gender Politics in Eighteenth-Century Natural History," *The American Historical Review* 98, no. 2 (1993): 382–411.

39 I thank Pete Langman for this phrasing.

known."[40] Such a perspective further complicates the notion of a static natural world that facilitated easy interpretation and categorisation.

For the purpose of clarity, the acts of collecting, describing, illustrating and classifying have here been discussed as a neat sequence. In practice, of course, and as we already saw, natural historical study was always a matter of intertwined, integrated activities: verbal species descriptions could be based on illustrations that were, in turn, made of preserved specimens. There was, therefore, usually not one specific entry point into the examination of a species, but multiple. How to weigh the relative values of these points of entrance was of great concern to naturalists throughout the eighteenth century. This book will explore these epistemic junctions.

4 Sources and Structure

The primary sources for answering the questions outlined above are natural historical works that describe and discuss fish, here called 'fish books'. These books were critical means by which observations and insights were communicated from one naturalist to another, and across generations as well as geographical regions. Although varying considerably in shape and form, the books that are examined here are fascinating as materialised efforts to capture fish on paper. This choice was a daring one. After all, these creatures went unseen most of their lives, submerged in the watery part of the world, making them more difficult to observe than most of the animals that lived on land. This strangeness of the element water may well have led to the 'terrestrial bias' that biologists observe today.[41]

The humanities also have to contend with this bias. Despite their significance in the history of science, the books examined here have hitherto seldom been studied – or else only recently. The titles of these three books (John Ray and Francis Willughby's *Historia piscium*, Peter Artedi's *Ichthyologia sive opera omnia de piscibus*, and, lastly, Marcus Élieser Bloch's *Allgemeine Naturgeschichte der Fische*) announce the breadth of their ambition: to examine the natural history of fishes in its entirety, rather than, for example, the fish of a particular region. The authors explicitly address why they have made certain decisions, and as such offer insight into their ideas about how natural

40 Whitney Barlow Robles, *Curious Species: How Animals Made Natural History* (New Haven: Yale University Press, 2023), 170.

41 Fred Van Dyke, *Conservation Biology: Foundations, Concepts, Applications* (Dordrecht, Springer: 2008), 313–314.

history should be carried out. Furthermore, this selection of three voluminous books, ranging as they do from several hundreds to several thousands of pages, allows for an in-depth study of the contents of these works, including the way in which they demarcate object and method, their processes of production, and their authors.

Each chapter takes as its place of departure the qualifications that the naturalists themselves use: how do they envision their work and its approach; and how do they evaluate their own contributions apropos those of earlier naturalists or their peers? At the same time, it is important to not take these qualifications at face value, but to try to understand why certain claims were made, and in which context they were made – and how they worked out in practice. Rather than looking at these authors through our contemporary lens, therefore, the historical actors are, as much as is possible, studied in their own context and on their own terms. This requires interrogating the terminology that naturalists themselves used, but also asking why they deployed these, and to what purposes and effects.

These fish books are the burnished outcomes of complex and sometimes messy research. Naturalists undertook field trips to examine living animals, visited cabinets of curiosities, purchased specimens and natural historical works at auctions, and consulted books in libraries. For naturalists, fishes needed to be caught, stored, preserved, circulated, described, classified and depicted – and not necessarily in that order. Besides the engagement of learned scholars, this entailed the involvement of multiple persons: from fishermen to fishmongers, cooks, artists, missionaries, pupils, publishers and printers. This is clear when considering not merely the end result, the three monographs in their published state, but also taking into account how these books were formed from their early beginnings. My examination of these works is therefore supplemented by manuscripts, travel accounts, circulars, letters, natural historical sketches and drawings and natural historical specimens.

Chapter 1 takes the definition of a fish as offered in the *Historia piscium* as its point of departure. This book, based on research undertaken by Willughby and Ray, was published under the auspices of the Royal Society in London. This chapter traces how the 'fish' as an object of study changed from being a member of the *aquatilia*, as it was up until the sixteenth century, to the creature with fins and without feet it had become by the late seventeenth, embedding this shift in changing attitudes towards the natural world. Willughby and Ray extracted fish from their element, but they also teased them out from the web of literary, encyclopaedic associations of which they had long formed part. In presenting a new footing for natural history, Willughby and Ray articulated ideals for description, classification and illustration based on direct observation, although in practice their work remained rooted in earlier traditions of natural history.

The second chapter takes the meanings of direct observation as its focus. In preparing the *Historia piscium*, Ray emphasised how they drew on their own observations, as well as those of friends and other reliable authorities. Despite their preference to see species with their own eyes, Willughby, Ray and their peers at the Royal Society simply could not examine each and every species for themselves. Fellows could turn to preserved specimens and illustrations, yet these came with their own epistemological uncertainties. To come to a clear understanding of species, direct observation of a (more or less) fresh fish was much preferred. Willughby and Ray took to ports and fish markets and made use of the 'experiential understanding' of fishermen and fishmongers. Such exchanges required negotiation and interpretation of the epistemological values that should be attached to experience.

Chapter 3 revolves around the relatively unknown Peter Artedi and his book *Ichthyologia, sive opera omnia de piscibus* (Leiden, 1738) published posthumously by his famous friend Linnaeus. He developed a strictly defined taxonomic system that encompassed classes, order, genera, species and finally, varieties – and drew up clear rules for establishing each. His book was intended as a programme for an 'ichthyology': Artedi drew up definitions for both the study of fish and its practitioners, and laid down the principles on which the latter should operate. He proposed a quantitative method for examining and describing fish rather than a qualitative one. In the process, he excluded any kind of knowledge that he considered 'amethodic', like the experience of fishermen, demarcating ichthyological knowledge from artisanal knowledge while also attempting to distinguish ichthyology from neighbouring fields such as comparative anatomy or other parts of zoology.

The fourth and final chapter zooms in on Marcus Élieser Bloch's *Allgemeine Naturgeschichte der Fische* (12 vols, Berlin, 1782–1795). This Jewish-German physician set out to complete the gaps in the classification systems of Artedi and Linnaeus. Finding even some of the most common fish in the Prussian states lacking, he began collecting fish from these states in his cabinet in Berlin. He soon expanded his collections to species from all over the world. This was made possible by a far-flung network of correspondents and collectors that rested on colonial infrastructures. Each of the species in his collection was described and depicted in his series of lavishly illustrated, handsomely hand-coloured books. The chapter argues how in the *Allgemeine Naturgeschichte der Fische*, Bloch pioneered an innovative format for the illustration of fish, creating a type of 'epistemic image' that combined those features of fish he found salient.

This made him an authority of the world's fish without having to leave Europe, or even travel far from Berlin.

Drawing these chapters together, the conclusion addresses the disciplination of natural history that took place over the course of the long eighteenth century. It shows how, even though the study of fish was an integrated branch of natural history until well into the eighteenth century, ichthyology came to be increasingly demarcated as a separate field with its own particular rules, and its own kind of practitioner. The conclusion first considers how a new method for classification served to separate ichthyological knowledge from practical, artisanal knowledge of fish, before examining how this process of disciplination shaped and was shaped by epistemic debates regarding the collecting, depicting and illustrating of fish.

Finally, the epilogue explores the opportunities that historical sources offer when it comes to our contemporary understanding of the world underwater. Whether we call them fish or not, the dazzling diversity of aquatic species at which the naturalists discussed in this book marvelled is under threat. The sources that are presented in this book contain a wealth of information when it comes to the biodiversity of the past. They comment on the quantity of species that dwell in certain waters, or expound on the quality of aquatic environments. Together, these sources can build up a picture of the occurrences of particular species through the centuries. As the epilogue outlines, this stands or falls with close collaboration between historians, biologists and ecologists as well as communities of practice. It requires us to comprehend the world underwater as a shared site of investigation, now and in the past.

CHAPTER 1

From *Aquatilia* to Fish

The *Historia piscium* opened with a clarification on the part of the authors regarding "what we understand under the name of Fish".[1] John Ray explained that they were:

> not ignorant of the fact that the name of *Fish* by some is extended as widely as possible to include all *Aquatilia*, both the ones which must be indicated as consisting of blood, and the bigger, bloodless ones, which *Aristotle* has divided in three genera, truly μαλάκια or *mollia* [soft], μαλακός[τ]ρακα or *crustacea* [with scales], όσρακοδερμα or *testacea* [covered with shells]. And indeed, our common people has all these for fish.[2]

He and his fellow author, the late Francis Willughby, however, departed from these definitions and introduced their own:

> But we shall use in this work the word for *fish* in the stricter sense, only for those aquatic animals lacking feet, covered either in scales or smooth skin, swimming by means of fins, living constantly in water, never of its own volition coming out onto dry land, and scarcely able to survive out of the water.[3]

The principal aim of this work, the outcome of years of sustained study he and Willughby had undertaken, was to offer the reader accurate descriptions of all known species of fish. Having investigated plants, insects, birds and fish together, they came to the conclusion that in studying living nature, clear demarcations were vital. This entailed not only distinguishing fish from

1 '*Quid Piscis nomine intellegimus*', *Hist. pisc.*, 1.

2 'Non sum nescius *Piscis* nomen à nonnullis quam latissime extendi ad *Aquatilia* omnia significanda tam sanguinea, quam exanguia majora, quae *Aristoteles* in tria genera dividit, nimirum μαλάκια seu *mollia*, μαλακόςτρακα seu *crustacea*, & όσρακοδερμα seu *testacea*. Quin & vulgus nostrum haec omnia pro piscibus habet.' Ibid.

3 'Verum nos in hoc opere restrictiore acceptione voce *piscis* utemur, pro iis tantum aquatilibus, quae & sanguinae sunt, & pinnis natant, & pedibus carent, & in aquis perpetuo degunt, ibidemque pariunt, nec unquam sponte in siccum exeunt, aut extra aquas diu vivere possunt.' Ibid.; extended translation of Kusukawa, "*Historia Piscium* (1686) and Its Sources," 308.

non-fish but also distinguishing one species from the other. And so they set out to put the study of natural history on a new, solid footing.

This chapter outlines the conceptual shift in seeing a fish as a creature that lives in the water (all *aquatilia*) to a creature without feet but with fins. Moreover, it examines this shift as it relates to the general changes of attitudes towards the study of nature seen as the sixteenth century moved into the seventeenth. To better understand Willughby and Ray's ambitions, it is worth considering the aims and scope of early modern natural history that had been passed down from the sixteenth century. This chapter highlights some of the important developments. It argues that, in presenting themselves as reformers of natural history, Willughby and Ray drew up new definitions and articulated an empirical program for observing and categorising species – but that their work was in effect a learned empiricism, as it remained rooted in earlier traditions of natural history with regard to observation, description and illustration. The ideals they espoused regarding description and illustration in particular serve to emphasise these ambitions. In the attention they paid to the physical characteristics that demarcated hippopotami from pike and carps from sharks, Willughby and Ray transformed the study of fish not only by epistemologically separating them from the element in which they dwelled, but also by formulating clear markers of species.

1 Different Ways to Define a Fish

Even though Ray tended to distance himself from earlier authors in his writings, their works remained as sources of knowledge of which both he and Willughby had made ample use. It was more a question of emphasis. Aristotle and Pliny the Elder were among the traditional authorities they cited, attesting to the lasting influence of the classical canon.[4] Engagement with these ancient works was, however, more of a critical act in the sixteenth century, as humanists became aware of the discrepancies between what they read in these texts and what they saw around them. While assertions of eye-witnessing had carried authority from antiquity, they took on new vigour in the early modern

4 Anthony Grafton, *New Worlds, Ancient Texts: The Power of Tradition and the Shock of Discovery* (Cambridge, Mass.: Harvard University Press, 1995), 248. They did so, however, on their own terms with their unique way of engaging with classical philosophical texts through an intricate textual, humanistic tradition, see: Dmitri Levitin, *Ancient Wisdom in the Age of the New Science: Histories of Philosophy in England, c. 1640–1700* (Cambridge: Cambridge University Press, 2015), 4.

period.[5] In the sixteenth century, a wide range of people – from artisans to anatomists and beyond – emphasised authority as derived from *autopsia*, or 'seeing with one's own eyes'.[6] Towards the end of the seventeenth century, the term *observatio* became firmly established as the vocabulary of experience.[7] Direct observation was now seen as a crucial tool for interpreting nature. This also had repercussions for how living beings were approached as a topic of study: no longer encapsulated in ancient texts that had been traditional sources of authority, they were now seen as possible proof for new philosophies that sought to reveal nature's underlying patterns. The shifting definition of fish fits into this development.

What did the term *aquatilia* encompass? Pliny the Elder had used the term to denote aquatic animals, as a substantive adjective for that which was "living, growing, or found, in or near water".[8] In the sixteenth century, the category of *aquatilia* encompassed all water-dwelling animals, from whales and rays to oysters, shrimp, crabs, turtles, walruses, octopuses (Figure 2) and so on. These are the species one encounters on the pages of, for instance, Belon's *De aquatilibus* (Paris, 1551) and Salviani's *Aquatilium animalium historiae* (Rome, 1554–58).[9] The title of Gessner's *Historiae animalium liber IIII. qui est de piscium & aquatilium animantium natura* (Zurich, 1558) mentions fish and *aquatilia* separately, but discussed them in one and the same work.[10] The Latin term Gessner used, *piscium*, a declension from *piscis* (fish), was used in scholarly circles to refer to those aquatic animals with scales.[11] But this usage

---

5 Gianna Pomata, "*Praxis Historialis*: The Uses of *Historia* in Early Modern Medicine," in *Historia: Empiricism and Erudition in Early Modern Europe*, eds. Gianna Pomata and Nancy G. Siraisi (Cambridge, Mass.: MIT Press, 2005), 113.

6 Smith, *Body of the Artisan*, 42.

7 Ibid. The term 'observation' could refer to the actual act of observing, or the (often written or drawn) report that was made of an observation. Dirk van Miert, "Introduction," in *Communicating Observations in Early Modern Letters (1500–1675): Epistolography and Epistemology in the Age of the Scientific Revolution*, ed. Dirk van Miert (London: The Warburg Institute, 2013), 3.

8 A Latin Dictionary, comp. Charlton T. Lewis and Charles Short (Oxford: Oxford University Press, 1879), s.v. *aquatilia*.

9 Pierre Belon, *De aquatilibus, libri duo: cum eiconibus ad vivam ipsorum effigiem, quoad ejus fieri potuit, expressis* (Paris: Carolum Stephanum, 1553); Hippolyto Salviani, *Aquatilium animalium historiae liber primus* (Rome: Hippolyto Salviani, 1554–1558). Another early example is Nicolaus Marschalk, *Historia aquatilium latine ac grace cum figuris* (Rostock: Nicolaus Marschalk, 1517).

10 Conrad Gessner, *Historiae animalium liber IIII. qui est de piscium & aquatilium animantium natura* (Zurich: Christoph Froschauer, 1558); henceforth cited as Hist. anim. IIII.

11 Florike Egmond, *Eye for Detail: Images of Plants and Animals in Art and Science 1500–1630* (London: Reaktion Books, 2017), 60.

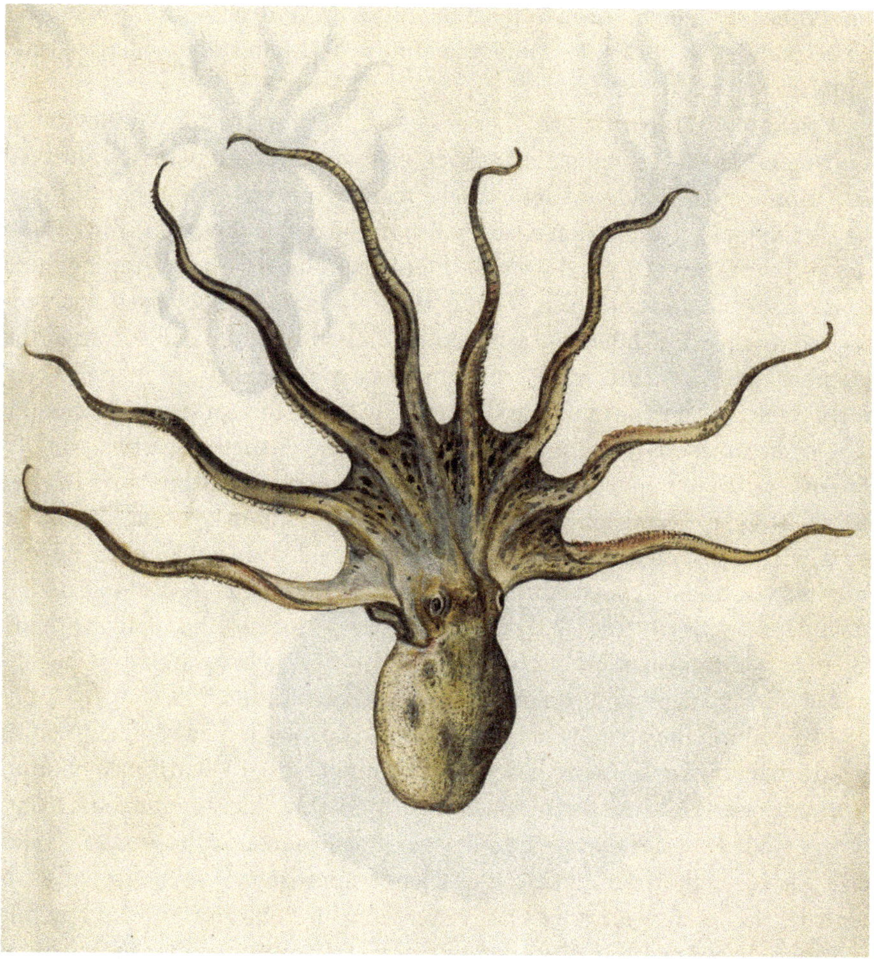

FIGURE 2 Watercolour of an octopus. Gessner-Platter Album, Hs. III C 22, 193
ALLARD PIERSON, UNIVERSITY OF AMSTERDAM

was not universal. In his *L'Histoire entière des poissons* (Lyon, 1558), Rondelet included both 'fish that do not have blood' (such as jellyfish) and 'fish with hard shells' (scallops, for example).[12] This use of the word fish corresponded to a broader tendency in early modern Europe for using the term for everything which lived either in, or on, the water.[13] This brief exploration of the term

12 'les poissons qui n'ont point de sang' and 'les poissons coverts [couverts] de test dur'; Guillaume Rondelet, *L'Histoire entière des poissons* (Lyon: Macé Bonhomme, 1558), table of contents.

13 Egmond, *Eye for Detail*, 60; Ogilvie, *Science of Describing*, 234–235.

underlines the anachronism implicit in only taking into account those species corresponding to our contemporary definition of fish in early modern natural histories of aquatic animals.

A defining feature of a fish, therefore, was that the creature dwelled in or on water. The correspondence between elements and animals was affirmed in Scripture: "And God said, let us make man in our image, after our likeness: and let them have dominion over the fish of the sea, and over the fowl of the air, and over the cattle, and over all the earth, and over every creeping thing that creepeth upon the earth."[14] Pliny the Elder had divided species according to whether they inhabited land, air or water;[15] Aristotle had accorded the elements (earth, air, fire, water) a central place in his theories of matter.[16] This elemental division continued to dictate daily life in the early modern period. The elements were thought to correspond to the changing seasons, the different ages of life, qualities (hot, dry, cold and moist) and humours (choler or yellow bile, blood, phlegm, and black bile). This may explain why, for sixteenth-century naturalists, *aquatilia* were considered a coherent group that merited treatment as such. Natural historical works on *aquatilia* of this period do indeed discuss the cold and moist properties of species, in keeping with the Galenic tradition, expounding at length on how these could either be beneficial or harmful depending on one's humoural constitution.[17]

As we saw in the opening of this chapter, Willughby and Ray defined 'fish' in a stricter, narrower sense than Gessner and other earlier Renaissance naturalists had done. While this definition of a fish is notably different from ours today, as it included species that we now categorise as sea mammals, such as whales and porpoises, their definition entailed a new approach to defining what a fish actually was (and what it was not). In a letter to Robinson in 1684, Ray furthermore explained that "[*e*]*xanguia aquatica* I account rather insects than fishes."[18] These *exanguia aquatica* or "bloodless aquatic creatures" included the shellfish and molluscs that, as we have seen, were counted among fish by other authors. For Willughby and Ray, what made a fish a fish was thus not that it swam in the water, but that it displayed specific physical characteristics.

14 Gen. 1:26.

15 Enenkel, "Die antike Vorgeschichte," 50.

16 See also: Eric Lewis, "Aristotle's Physical Theory," in *The Cambridge History of Science*, eds. Alexander Jones and Liba Taub, vol. 1, *Ancient Science* (Cambridge: Cambridge University Press, 2018), 196–214.

17 David Gentilcore, *Food and Health in Early Modern Europe: Diet, Medicine and Society, 1450–1800* (London: Bloomsbury, 2016), 19.

18 Ray to Robinson, 13 March 1684 (OS), *The Further Correspondence of Ray*, ed. Robert W. Theodore Gunther (London: The Ray Society, 1928), 164.

Some decades before the *Historia piscium* was published, the philosopher and mathematician René Descartes (1596–1650) hypothesised that nature was not structured from the four elements, but from small particles that clashed and collided with one another. He wished to explode the Aristotelian system and replace it by his own, mechanist one.[19] In his *Principia philosophiae* (Amsterdam, 1644), Descartes offered a new metaphysical framework for the study of planetary and celestial motions, the inner structures of the earth, as well as the physiologies of plants and animals.[20] All creatures could be studied on a par with one another, with every creature relegated to the same ontological status, *viz.* that of a machine that was structured and governed by the fixed laws of nature.[21] One could understand the working of these laws through a close examination of a flea, a flounder or a falcon – it did not matter, for they were all made of the same matter that followed the same laws. By implication, the erstwhile organisation of animals into elemental realms lost its relevance. Although Willughby and Ray did not make the philosophical underpinnings of their new definition explicit, they were well aware of Descartes' ideas. Willughby dedicated several pages of notes to Cartesian philosophy in his commonplace book, while Ray purchased a copy of the *Principia philosophiae.*[22]

Ray's treatment of the spawning of fish neatly embodies one pressing topic in which classical notions failed to match the new philosophies: that of procreation. He summarised how Aristotle had rightly discerned three different types of fish, each of which had their own mode of propagation: first, the cetaceans that reproduced in the same manner as mammals did, by giving birth to live young and were thus viviparous; second, the cartilaginous fish that were ovo-viviparous, hatching eggs in their own body, and then bringing forth live young; and, lastly, the spiny fish that were oviparous, which meant that the females laid eggs (called roe) which were then fertilised by the milt (seminal fluid) of males.[23] Ray also cited Aristotle's idea that some species of fish were neither oviparous nor viviparous, generating spontaneously from mud, sand

19 Eric Jorink, "Insects, Philosophy and the Microscope," in Curry, Jardine, Secord, and Spary, *Worlds of Natural History*, 136.

20 A systematic discussion of Descartes' *Principia philosophiae* is offered in Stephen Gaukroger, *Descartes' System of Natural Philosophy* (Cambridge: Cambridge University Press, 2002).

21 Jorink, "Insects, Philosophy and the Microscope," 136.

22 Richard Serjeantson, "The Education of Francis Willughby," in *Virtuoso by Nature: The Scientific Worlds of Francis Willughby* (FRS), ed. Tim Birkhead (Leiden: Brill, 2016), 75, 94–95, 97.

23 *Hist. pisc.*, 16–18.

or foam.[24] The theory of spontaneous generation was broadly agreed upon by sixteenth-century naturalists. Rondelet even made matters more complex by contending that the carp's procreative habits were ambiguous, as it could sometimes proceed from eggs, and at other times from mud and sand.[25]

By the second half of the seventeenth century, however, this assumption had become more problematic. On the one hand, spontaneous generation accorded with the everyday experience of seeing living beings emerge from lifeless matter, like worms that appeared in rotten food. On the other hand, the element of chance that this process entailed seemed to elide the fixed laws of nature as postulated by Descartes.[26] Naturalists began to seek explanations that rendered the process of generation regular and predictable. Ray included these explanations into his discussion, like the claim of William Harvey (1578–1657) that all living beings came from eggs.[27] He tellingly closed his discussion with a reference to the findings of a certain 'D. Levenhoeck', who had observed *animalculorum* [literally: little animals, spermatozoa] in the seminal fluid of fish.[28] This was, of course, the Delft cloth merchant Antoni van Leeuwenhoek (1632–1723), who had been corresponding with the Fellows of the Royal Society since 1673 and sent them drawings of what he observed through his lenses.[29] He took a keen interest in the procreation of underwater creatures, even the smallest ones, examining the seed of pike, cod and oysters (sourced from local fishermen) under his microscope.[30] Ray admitted that the

---

24 Aristotle, *History of Animals in Ten Books*, book VI, chapter XIV, ed. and trans. Cresswell, 157.

25 Rondelet, *L'Histoire entière des poissons*, 108. In the loose and abbreviated German translation that Conrad Forrer (*c.*1530–1594) made of Gessner's *Hist. anim. IIII*, Rondelet's statement that carp are sometimes born from chaos and dirt, and sometimes from seed and roe is copied, see: Conrad Gessner, *Fischbuch*, trans. Conrad Forrer (Zurich: Christopher Froschauer, 1563), 164–165.

26 Jorink,"Beyond the Lines of Apelles," 153–159. Descartes himself did not entirely rule out spontaneous generation, see: Eric Jorink, "Snakes, Fungi and Insects. Otto Marseus van Schrieck, Johannes Swammerdam and the Theory of Spontaneous Generation," in Smith and Enenkel, *Zoology in Early Modern Culture*, 199–207.

27 This was William Harvey's *Exercitationes de generatione animalium* (London: Du Gardianis, 1651).

28 *Hist. pisc.*, 18.

29 Lodewijk Palm, "Leeuwenhoek and Other Dutch Correspondents of the Royal Society," *Notes and Records of the Royal Society of London* 43, no. 2 (1989): 192–193; Sietske Fransen, "Antoni van Leeuwenhoek, His Images and Draughtsmen," *Perspectives on Science* 27, no. 3 (2019): 485–544.

30 Marlise Rijks, "A Taste for Fish: Paintings of Aquatic Animals in the Low Countries (1560–1729)," in Paul J. Smith and Florike Egmond (eds.), *Ichthyology in Context (1500–1880)* (Leiden: Brill, 2023), 270–271; Edward G. Ruestow, "Images and Ideas: Leeuwenhoek's Perception of the Spermatozoa," *Journal of the History of Biology* 16, no. 2 (1983): 188, 199.

matter of spontaneous generation remained a difficult question, but it is clear that he followed the discussion closely.[31]

By the later seventeenth century, ancient works had been largely superseded by contemporary ones, and classical claims were no longer at the forefront of research agendas.[32] New philosophies challenged hitherto widely accepted views on the governing structures of nature. The hierarchical view of nature gradually gave way to a more mechanist one, in which the anatomy and physiology of each and every animal was governed by the same natural laws. Salient in this regard is how the definition of fish given in the *Historia piscium*, when compared to those earlier books on which it drew, relied on its physical characteristics (namely, having fins and lacking feet), rather than on its surroundings (namely, living in the water). We will now turn to examine the *Historia piscium* in more detail, from its earliest inceptions to its published form.

2 "Useful Studies and Designs"

When Willughby passed away after a short illness in 1672, his death came in the midst of "the hot pursuit of useful studies and designs", as Ray put it.[33] Willughby had been a polymath, well versed not only in Cartesian philosophy, mathematics, chymystry and magnetism, but also in the study of plants, four-footed beasts, birds and fish.[34] In Ray he had found a kindred spirit. Ray tutored in Greek and mathematics, but came to prefer the study of natural history, which he considered as a perfect entry into contemplating God's endless wisdom.[35] Both men immersed themselves in various branches of learning, and their interests stretched from the movement of the heavens to classical languages, to fish.

---

31 *Hist. pisc.*, 18. For more on Willughby and Ray's attitudes towards spontaneous generation see: Brian Ogilvie, "Insects in John Ray's Natural History and Natural Theology," in Enenkel and Smith, *Zoology in Early Modern Culture*, 235–262; Ogilvie, "Willughby on Insects," 350–351. John Ray's *Historia insectorum* (London: A. & J. Churchill, 1710) appeared posthumously.

32 Ogilvie, "Visions of Ancient Natural History," 25.

33 Willughby, *Ornithology*, preface, sig. A2v.

34 He studied so much that his tutor thought 'he should moderate his thirst for learning', see: William Poole, "The Willughby Library in the Time of Francis the Naturalist," in Birkhead, *Virtuoso by Nature*, 236.

35 Ray's most influential physico-theological treatise, *The Wisdom of God Manifested in the Works of the Creation* (London: Samuel Smith, 1691) was reprinted in 1692, 1701 and 1704; it was translated in German, Dutch and French.

The basis for the collaboration of Willughby and Ray was formed at the University of Cambridge. In 1644, Ray had been admitted to Trinity College on a stipend. After completing his degree, he stayed on as a lecturer and tutor, first of Greek and later in mathematics. During this time, he developed a botanical interest, cataloguing the plants in the vicinity of Cambridge, and cultivating them in his garden.[36] Willughby, who came from a noble family, arrived at Trinity in 1652.[37] He enrolled in the undergraduate arts course, and although it was uncommon for gentlemen to stay at the university for an extended period of time, he went on to take a Master of Arts, all the while cultivating a particular interest in natural philosophy and natural history. It was likely in this context that he got to know Ray better; the latter's *Catalogus plantarum circa Cantabrigiam nascentium* (Cambridge, 1660) mentions him as a contributor.[38]

In the early 1660s, both Willughby and Ray departed from Cambridge.[39] They continued to meet fellow scholars at London's Royal Society for the improving of natural knowledge in London. This had been established in 1660 under a charter of King Charles II (1630–1685) and foregrounded direct experience as the foundation for acquiring knowledge about nature. Given their sustained interest in studying the natural world, Willughby and Ray seemed suitable members, and they knew several of the other Fellows from their days at Cambridge. The Society's meetings offered a congenial setting for men of certain standing interested in natural philosophy to exchange thoughts, ideas, and occasionally to perform experiments.[40]

Their approach is often seen as explicitly empirical; the Society's motto, *nullius in verba* (on the word of no one), reflects the commitment of the Fellows to questioning received wisdom, and to examining everything with their own eyes.[41] The chronicler of the early Royal Society, Thomas Sprat (1635–1713)

---

36 For Ray's years at Cambridge, see: Raven, *John Ray*, 21–110.

37 For the history of the Willughby family, see: Cassandra Willoughby, *An Account of an Elizabethan Family: The Willoughbys of Wollaton by Cassandra Willoughby, 1670–1735*, ed. Jo Ann Hoeppner Moran Cruz in *Royal Historical Society Camden Fifth Series* 55 (2018): 67–258.

38 Willughby's time at Cambridge is discussed in more detail in Serjeantson, "The Education of Francis Willughby," 44–98.

39 Raven, *John Ray*, 60–61.

40 A concise overview of early modern academies is given in Jürgen Renn and Florian Schmaltz, "Institutions and Knowledge Systems: Theoretical Perspectives," in *The Institutionalization of Science in Early Modern Europe*, eds. Mordechai Feingold and Giulia Giannini (Leiden: Brill, 2019), 292–296.

41 That experience was not to be equated with objectivity is argued in Alexander Wragge-Morley, *Aesthetic Science: Representing Nature in the Royal Society of London, 1650–1720* (Chicago: Chicago University Press, 2020), 3–4.

wrote that the Fellows took: "[...] as their *Fundamental Law*, that whenever they could possibly get to *handle* the subject, the *Experiment* was still perform'd by some of the *Members* themselves."[42] They collected many observations on a wide range of topics from which they hoped to distil general natural principles. In this, the Society owed much to the philosophical programme of Francis Bacon (1561–1626), even if its adherence to the latter's work was not absolute. Chapter 2 will delve into the historiography surrounding the Royal Society in much more detail than the present chapter allows, and look at its empirical project in more depth.

Willughby and Ray spent the larger part of this decade travelling. During their time at Cambridge, they made several trips within England and Wales collecting observations of plants, birds and fish as well as compiling lists of words in several dialects.[43] Between 1663 and 1666, Willughby and Ray toured continental Europe with fellow Cantabrigians Nathaniel Bacon (1647–1676) and Philip Skippon (1641–1691) as well as two servants. After having reached Calais, the company set course to the Low Countries, from where they travelled through Germany, Switzerland, and Austria to the Italian Peninsula – after which the company returned home.[44] Ray's and Skippon's travel accounts demonstrate the group's devotion to recording everything they encountered, a devotion remarkable in its intensity, attention to detail and scope.[45] Alongside animals and plants, the naturalists documented inscriptions on buildings and local customs. Skippon wrote, for example, that during their visit to Rome, they had climbed Trajan's Column, noting that it comprised 173 steps.[46] With this, they proved that the humanist scholar Alfonso Chacón (1530–1599) statement that the column had 184 steps as incorrect.[47] While this example might seem

42 Sprat, Thomas. *The History of the Royal-Society of London for the Improving of Natural Knowledge* (London: John Martyn, 1667), 83.

43 Sachiko Kusukawa, "The *Historia Piscium* (1686)," *Notes and Records of the Royal Society of London* 54, no. 2 (2000): 179.

44 Except for Willughby, who ventured onwards to Spain. For a more detailed itinerary, see: Mark Greengrass, Daisy Hildyard, Christopher D. Preston and Paul J. Smith, "Science on the Move: Francis Willughby's Expeditions," in Birkhead, *Virtuoso by Nature*, 152–193.

45 John Ray, *Observations Topographical, Moral and Physiological, Made in a Journey through Part of the Low-Countries, Germany, Italy and France* (London: John Martyn, 1673) and Skippon, *Journey* cited note 28; Willughby's account of the journey unfortunately appears no longer extant, though parts of it were subsumed in Ray's. Greengrass, Hildyard, Preston and Smith, "Science on the Move," 163.

46 Skippon, *Journey*, 653.

47 Alfonso Chacón, *Historia iutruisque belli datici a Traiano Caesare gesti quae in columna euisdem Romae visuntur, collecta* (Rome: Franciscum Zanettum & Bartholomaeum Tosium socios, 1576), A1r.

trivial, it is exemplary of their insistence on holding their own direct observations against the reports of earlier authors.

Their travels both within the British Isles and on the Continent were important opportunities for the examination of nature. In Wales and the West Country, for example, they described birds, fishes and plants as well as visiting mines.[48] On the Continent, they visited the cabinet of curiosities of the late Felix Platter (1536–1614) in Basel.[49] They saw a "[...] collection of rarities; among which [...] many sorts of minerals, stones, dry'd fishes, *&c.* with their names written; a lamp with a brass globe, which, turned any way, would still keep in its right posture; lachrymal urns; painted books of quadrupeds, fishes and fowls; [...]."[50] It was in the natural historical specimens in such collections that Willughby and Ray also caught glimpses of species not native to Europe. They could make use, for example, of the Royal Society's Repository, which had drawn "together in one Room, the greatest part of all the several kinds of things, that are scatter'd throughout the *Universe*."[51] The catalogue that natural historian and Fellow of the Royal Society Nehemiah Grew (1641–1712) drew up of this collection gives an idea of its contents, ranging from coins to 'humane rarities' and to animals, plants and minerals.[52]

Upon returning to England, Ray was invited to take up residence at the Willughby family's estate, Middleton Hall in Warwickshire.[53] This meant that they could all the more easily pursue their shared and ambitious goal: to identify and describe all the species of plants, birds, fish and insects they had come across, and to put them into some sort of order. This arrangement continued until Willughby's premature death in 1672. Ray continued his work, turning the

48 Greengrass, Hildyard, Preston and Smith, "Science on the Move," 151.

49 Skippon, *Journey*, 446.

50 Philip Skippon, *An Account of a Journey Made Thro. Part of the Low Countries, Germany, Italy and France* (London: Churchill, 1732), 446; the painted books might have been what are now known as the Gessner-Platter albums held in the Special Collections Department of the Universiteitsbibliotheek Amsterdam (hereafter UBA), Amsterdam, hs. III C 22/23; for its recent discovery and its provenance, see Florike Egmond, "A Collection within a Collection: Rediscovered Animal Drawings from the Collections of Conrad Gessner and Felix Platter," *Journal for the History of Collections* 25, no. 2 (2013): 149–170.

51 Sprat, *The History of the Royal-Society*, 251.

52 Nehemiah Grew, *Musaeum Regalis Societatis, or, A catalogue & description of the natural and artificial rarities belonging to the Royal Society and preserved at Gresham Colledge* (London: Hugh Newman, 1681), 1–10.

53 Dorothy Johnston, "The Life and Domestic Context of Francis Willughby," in Birkhead, *Virtuoso by Nature*, 7.

extensive notes he and Willughby had produced into actual natural historical manuscripts that were fit for print.[54]

In 1685, more than ten years after Ray had commenced his preparation of the history of fishes, the Royal Society received word that it was almost ready. As is the case for the *Ornithologia*, the book that Willughby and Ray delivered on birds, the *Historia piscium* displays Willughby's name on its title page as the author.[55] Ray, who had assembled the existing materials, supplementing them where necessary, was credited as the work's editor. The state of the work at the moment of Willughby's death and the decade it took Ray to convert their notes into something publishable has led to much debate regarding which of the two men ought truly be considered the book's author.[56] In his biography of Ray, Charles Raven more than once contends that the contributions of the younger, less experienced Willughby to their natural historical researches could only have been minor.[57] Two recent publications that focus solely on Willughby have done much to bring to light the latter's significant role in the duo's fruitful cooperation.[58] Their contributions can, to some extent, be teased apart. The *Historia piscium* contains a few passages where Ray explicitly attributes a certain statement to Willughby, and the title page specifies that the first two parts of their publication came from Ray's hand. On the whole, the history of fishes is best considered as a collaborative project to which Fellows other than its prime movers, Willughby and Ray, also contributed.

The most thorough study of the *Historia piscium*, both with regard to its intellectual underpinnings and the more practical side of publication, has been undertaken by Sachiko Kusukawa.[59] As Kusukawa has stated, the Fellows' col-

54 In this, he was hindered by no longer having access to Willughby's notes at Middleton. It is unclear why he had not copied these notes: there may have simply been too many, or he might not have expected to lose access to them in the first place.

55 Francis Willughby and John Ray, *Francisci Willughbeii Armig. De historia piscium libri quatuor* (Oxford: Sheldonian Theatre, 1686), title page. This work will henceforth be cited as *Hist. pisc.*

56 This debate is addressed in more detail in Isabelle Charmantier, Dorothy Johnston and Paul J. Smith, "The Legacies of Francis Willughby," in Birkhead, *Virtuoso by Nature*, 382–385.

57 Raven, *John Ray*, 51.

58 Birkhead, *Virtuoso by Nature*, and Birkhead, *The Wonderful Mr. Willughby: The First True Ornithologist* (London: Bloomsbury, 2018). Willughby is often counted among the *virtuosi*; for discussions of the category of the *virtuoso*, see the special issue *The Varied Role of the Amateur in Early Modern Europe* edited by Pamela Smith in *Nuncius* 31, no. 3 (2016): 485–609.

59 Kusukawa, "The *Historia Piscium* (1686)," 179–197; Sachiko Kusukawa, "*Historia Piscium* (1686) and Its Sources," in Birkhead, *Virtuoso by Nature*, 305–334. An account of the

lective engagement with the work fundamentally shaped the way it was finally published.[60] The Royal Society became involved in the production of the work after they had found that Ray wanted to supply it with illustrations, as he had done for the *Ornithologia*. It proved difficult to find a printer willing to publish such a work because of the high costs involved. Eager to see the work in print, the Society assumed financial responsibility for its publication,[61] having previous experience in licensing and sponsoring books.[62] The Society's involvement was not just of financial nature: the Fellows helped to amass relevant material for the book and evaluated which observations merited inclusion. They also passed down their own observations.[63] Tancred Robinson (1658–1748) and Martin Lister (1639–1712) were especially active in delivering the *Historia piscium* to publication.[64]

When the book finally made it into print in 1686, it comprised four parts. It opened with general discussion of fish and their properties, continued with descriptions of cetaceans (whales and the like), descriptions of fish with cartilaginous skeletons (for example, rays and sharks), and lastly with descriptions of fish that had bony skeletons and spiny fins, like herrings and trouts, which made up the largest group. As will be dicussed in more detail below, such distinctions harked back to Aristotle, who Willughby explained had rightly grouped fishes into *cetacei* (whales), *cartilaginei* (cartilaginous fish) and *spinosi* (bony fish).[65] The appendix of the book contained descriptions of fish found in other books that were not considered precise enough to warrant inclusion in the body text.[66] The resulting work was a voluminous and impressive work in folio format.

---

publication process is also given in Anna Marie Roos, *Web of Nature: Martin Lister (1639–1712): The First Arachnologist* (Leiden: Brill, 2011), 318–332.

60 Kusukawa, "The *Historia Piscium* (1686)," 180.

61 Sachiko Kusukawa, "*Historia Piscium* (1686) and Its Sources," 305.

62 Tara Nummedal and Paula Findlen, "Words of Nature: Scientific Books in the Seventeenth Century," in *Thorton and Tully's Scientific Books, Libraries and Collectors: A Study of Bibliography and the Book Trade in Relation to the History of Science*, ed. Andrew Hunter (Aldershot: Ashgate, 2000), 189–192.

63 These letters were written in English, and excerpts of them were translated into Latin for inclusion in the *Historia piscium*, for example the experiment for determining the centres of gravity for a pilchard and a herring by holding the specimens by the tip of their back-fin: Tancred Robinson to Ray, 8 September 1685 (OS), *The Correspondence of John Ray*, ed. Edwin Lankester (London: The Ray Society, 1848), 174; cf. *Hist. pisc.*, 224.

64 For Lister's contributions, see: Roos, *Web of Nature*, 318–332.

65 *Hist. pisc.*, 21.

66 These included descriptions by Lister as well as from Nieuhof's *Gedenkweerdige Brasiliaense Zee- en Lant-Reize*.

In addition to its well over 300 pages of text, *Historia piscium* contained 189 sumptuous, often full-page, copperplate engravings, depicting no fewer than 388 species. The book was, in fact, originally envisioned as two separate works: one work containing texts, and the other (which was to bear the name *Icthyographia*) illustrations. While the works were eventually published together as one book, both parts retained their own title page. In most copies of the *Historia piscium* the illustrations are bound together, rather than being interleaved between the descriptions.[67] Because copper plates were expensive, Fellows and other donors subscribed to them; their names are inscribed on the plates. This brought down the cost by almost £163, a sum that did not match the final production cost of £360, which included the commissions to the engravers, the cost of the paper, and the printing of index and text.[68] One could purchase a copy for a little over £1 on a lesser quality paper if one had subscribed, or spend £1 and 8 shilling for the best paper if one had not.[69] Sales, however, proved disappointing and the Society suffered a considerable loss. Publishing books was often a risky affair, but its many engravings made the *Historia piscium* a particularly delicate enterprise.[70] Its publication turned out to be such a costly venture that the Society famously did not have sufficient funds left with which to finance the publication of Isaac Newton's *Philosophiae naturalis principia mathematica*.[71]

Those today who are surprised that a book of fish almost sank Newton's *Principia*, besides having the gift of hindsight, overlook the fact that a natural history of fish fitted snugly within the Royal Society's explicit programme. After all, the Fellows were invested in bringing together observations on a wide range of topics; Robert Hooke (1635–1703) insisted that attention should be paid to common things.[72] The minutes of the Society's meetings certainly reflect their investment in recording a wide range of phenomena in detail:

---

67 See also: Adrian Johns, *The Nature of the Book: Print and Knowledge in the Making* (Chicago: University of Chicago Press, 2000), 489.

68 Kusukawa, "The *Historia Piscium* (1686)," 191.

69 Ibid.

70 A broad treatment of the ways in which money was made (or lost) in the business of books in the early modern period can be found in: Shanti Graheli, ed., *Buying and Selling: The Business of Books in Early Modern Europe* (Leiden: Brill, 2016).

71 Kusukawa, "The *Historia Piscium* (1686)," 193; after the *Historia piscium*, the Society would refrain from offering direct financial support of projects, see: Nummedal and Findlen, "Words of Nature," 191.

72 Felicity Henderson, "Robert Hooke and the Visual World of the Early Royal Society," *Perspectives on Science* 27, no. 3 (2019): 398.

from the workings of quicksilver to the colour variants of the teeth of sheep.[73] The communications that were published in the *Philosophical Transactions* do likewise.[74] In short, in their shared, "hot pursuit" of useful studies, the Fellows did not regard scrutinising plants, birds, fish and insects in all their intricacies as an endeavour intrinsically less worthwhile than studying the motions of planets.

3 Formats for Description

In the epilogue of the history of fishes, Ray explained that it was not meant as comprehensive account of everything that had ever been written about all fish. That, he stated, had already been done by the great Gessner.[75] Instead, they included only those observations made by themselves, their friends, and other trustworthy witnesses. Through careful observations, they sought to set right the multiplication of species that earlier authors of natural history had caused by describing two or three species where there had, in fact, only been one. In keeping with their ambition to offer a clear demarcation of species they designed the book's species descriptions to facilitate easy identification. Their new approach to description hinged on a clear (though not exclusive) focus on the physical characteristics of a species put forth in a structured manner, and eschewed the cultural or literary context of the species at hand.

Description had been an important part of *historia naturalis* since its inception, but truly became a topic of concern in the sixteenth century.[76] Naturalists were confronted with the need to draw up species descriptions in such a way that it would be clear to their correspondents as well as the readers of their books precisely what they had seen. They examined plants and animals for differences [*differentiae*] in areas such as the shape or placement of leaves or petals, through which one species could be distinguished from another. These physical descriptions were usually combined with excerpts from literature, poetry, proverbs, myths, fables and emblems in which the species at hand

73 Thomas Birch's *History of the Royal Society*, 4 vols (London: A. Millar, 1756–1757) offers accounts, albeit necessarily abridged and edited, of what transpired at the meetings of the Royal Society. Birch, *History of the Royal Society*, vol. 1, 20 and vol. 3, 97.

74 On the early history of the *Philosophical Transactions* see: Noah Moxham, "Authors, Editors and Newsmongers: Form and Genre in the Philosophical Transactions under Henry Oldenburg," in *News Networks in Early Modern Europe*, eds. Joad Raymond and Noah Moxham (Leiden: Brill, 2016), 465–492.

75 *Hist. pisc. app.*, sig. Hr/v. Kusukawa, "The *Historia Piscium* (1686)," 182.

76 Ogilvie, *Science of Describing*, 6.

figured – in short, its cultural context.[77] This web of associations helped to make sense of the species and its place in the world.

Renaissance authors used various principles for grouping species. Species descriptions could be sequenced based on medicinal uses, habitat – Rondelet, for example, sorted fishes according to whether they swam in the sea, in rivers, in lakes or in marshes – or whether species were common or rare.[78] Other categorisations were based on morphological and physiological characteristics, and more closely resemble contemporary ideas of taxonomy.[79] Gessner, for example, sorted physically similar species of fish into groups, such as 'trout-like' species which, among others features, shared an adipose fin.[80] The Swiss naturalist also explained how one might tell one species from another. In the group of 'herring-like' fish, for example, the sardine could be distinguished from the herring because, while they might look much alike, the former was smaller.[81] The order in which he discussed all species throughout his *Historia animalium* was alphabetical: the trout- and herring-like species were not placed in proximity to one another on the pages of his work. To compare morphologically similar species the reader had to browse back and forth through the book. In his *Nomenclator aquatilium animantium* of 1560, Gessner forewent the alphabetical arrangement used in his *Historia animalium*. He first sorted fish into marine and fresh water species and then into categories based on physical properties, such as whether they were scaly, broad and flat or cartilaginous and long, echoing Aristotle.[82] It is indicative of how, in their publications, Renaissance naturalists drew on different ordering principles, either environmental (marine or fresh water) or morphological features (shape and skeletal structure) – or a combination of these.

As mentioned, Willughby and Ray's *Historia piscium* also adhered to the three overarching groups that Aristotle had established for the finny tribe: cetaceans, cartilaginous fish, and spiny fish, each of which were relegated to

---

77 Charmantier, "Emblematics in Ornithology in the Sixteenth and Seventeenth Centuries," *Emblematica* 18 (2010), 84.

78 Ogilvie, *Science of Describing*, 216.

79 For a discussion of different forms of taxonomy, see Ogilvie, *Science of Describing*, 219–228.

80 Sophia Hendrikx, "Gessner's Taxonomical Skill Exhibited in his Discussion of Felchen," in *Conrad Gessner (1516–1565) Die Renaissance der Wissenschaften/The Renaissance of Learning*, eds. Urs B. Leu and Peter Opitz (Zurich: De Gruyter Oldenbourg, 2019), 610.

81 Sophia Hendrikx, "Identification of Herring Species in Conrad Gessner's Ichthyological Works: A Case Study on Taxonomy, Nomenclature, and Animal Depiction in the Sixteenth Century," in Enenkel and Smith, *Zoology in Early Modern Culture*, 158.

82 Sophia Hendrikx, *Conrad Gessner's Fish Books (1556–1560): Processing Information in a Rapidly Expanding Field of Knowledge* (PhD diss., Leiden University, 2024), 70–73.

their own part of the *Historia piscium*. The cartilaginous fish was further split it in different subgroups largely based on shape (e.g., whether a species was long, flat, thick or thin).[83] These groups were, in turn, divided into smaller groups according, for example, whether or not they had teeth.[84] In their categorisation of spiny fish, shape again played a role, as did characteristics such as the number of dorsal fins and whether or not the rays in the fins were soft or thorny.[85] All species descriptions were sequenced according to this order. Although the book does arrange species into groups, their demarcations are relatively loose and ought not be confused with the three-tiered taxonomical hierarchy of order, genus and species that later naturalists established as will be discussed in Chapter 3.

It was the species themselves that were the prime concern for Willughby and Ray.[86] By observing species with scrupulous attention to detail, they aimed to uncover the 'true' (*viz.* God given) arrangement of species. The great multitude of species, Ray argued, had been made to "manifest and display the Riches of the Power and Wisdom of God."[87] As he contended in the preface to the *Ornithology*, earlier descriptions of birds had been "in many particulars confused and obscrue [obscure]",[88] but naturalists could dispel the confusion and multiplications of species by establishing 'characteristic marks' [*notae characteristicae*] on the basis of which species could be demarcated.[89] With their concern for characteristic marks, Willughby and Ray ensured that "the Reader might be sure of our meaning, and upon comparing any Bird with our description not fail discerning whether it be the described or no."[90] For fish, the characteristic marks might be the number and position of fins, certain spots or colours, or other properties – as long as these properties could be inferred from the actual specimen itself. Take, for instance, the sharks. They were discussed in the third part (on cartilaginous fish) and placed under the header of 'long and cartilaginous' fish.[91] In this group, the tope shark could be discerned from

83 *Hist. pisc.*, 46.
84 Ibid.
85 A detailed overview of the morphological groupings of the *Hist. pisc.* can be found in Kusukawa, "*Historia Piscium* (1686) and Its Sources," 309.
86 Ray later summarised and developed their methods for classifying fish in *Synopsis methodica avium & piscium* (London: William Derham, 1713).
87 Ray, *The Wisdom of God*, 369.
88 Willughby, *Ornithology*, preface, sig. A4v.
89 The notion of "characteristic marks" is explained in more detail in Kusukawa, "The *Historia Piscium* (1686)," 182.
90 Willughby, *Ornithology*, preface, sig. A4v.
91 *Hist. pisc.*, 47–64.

the similarly looking smooth hound shark by its larger size, its rows of sharp teeth and its eyes, the irises of which were of a brighter, silver colour.[92]

Their species descriptions follow the same general format. Having given the name(s) of the species at hand, they dived right into detailed descriptions of the outer and inner parts of the fish. Topics covered included, among other things, the size of the fish, its shape, colour, its head, its teeth, its scales, and its fins. In quite a few instances, they offered the measurements of a specimen, or wrote down the number of teeth or fin rays that they counted from it – a practice that, as we will see in the following chapters, would gain considerable currency among later naturalists. Willughby and Ray did not stop short of describing physical properties and made the occasional remark on the degustatory qualities of fish, commending the flesh of the herring, for example, as fat, soft and delightful.[93] In order to guide the reader through all these minute descriptions, key words were printed in the margin of the species descriptions which signalled the topic under discussion in the adjacent text: from the manner in which a species procreated to a description of its spleen.[94]

In describing the inner parts of fish, writers could take recourse to the accounts published by sixteenth-century naturalists, who routinely dissected animals.[95] Rondelet in particular, with his background in comparative anatomy, was known for performing dissections of fish, and Willughby and Ray's *Historia piscium* often cites him on these.[96] But, in keeping with the authoritative weight attached to an autopsy, Willughby and Ray would cut open specimens themselves if they had the chance. Their travels gave them ample chance to do so, possibly in the comfort of their lodgings.[97] A unique insight into this process is offered by a set of four drawings in the Middleton collection in Nottingham which show the step-by-step dissection of a male flair (a species of ray) that took place under the auspices of their travel companion Philip Skippon; the images are inscribed with Skippon's notes.[98] The first

---

92 Cf. descriptions of 'Mustelus laevis secundus' in *Hist. pisc.*, 51 and 'Mustelus laevis primus', ibid., 60.

93 *Hist. pisc.*, 219.

94 For this textual organisation tool, see: Ann Blair, "Annotating and Indexing Natural History," in *Books and the Sciences in History*, eds. Nicholas Jardine and Marina Frasca-Spada (Cambridge: Cambridge University Press, 2000), 72.

95 See: Anita Guerrini, *The Courtiers' Anatomists: Animals and Humans in Louis XIV's Paris* (Chicago: Chicago University Press, 2016), 57–61.

96 See also: Gillian Lewis, "The Debt of John Ray and Martin Lister to Guillaume Rondelet of Montpellier," *Notes and Records of the Royal Society of London* 66, no. 4 (2012): 323–339.

97 Kusukawa, "*Historia Piscium* (1686) and Its Sources," 316.

98 Drawings of the dissection of a flair, Middleton Collection (hereafter Mi LM), Special Collections Department of University of Nottingham (hereafter NUL), Mi LM 25/12–15; drawings 14 and 15 are reproduced in Birkhead, *The Wonderful Mr. Willughby*, 120.

image shows the contours of the flair drawn with lead.[99] The emphasis, however, is on the flap of skin that has been folded open to show the fish's insides. For this part, drawing ink is used – making it clear what part of the image is background and what is foreground. As the marginalia indicate, the flair's gallbladder has been made visible by removing the liver. In what is presumably the final image in this series (Figure 3), the skin of different parts of the body has been sliced open, its thorax and various organs laid bare. The heart might have been revealed, Skippon noted, were it not for the fact that the 'workman', Mr Okely, grew tired and wished to finish the drawing and hand it over.[100] The precise recording of this process, though subject to limitations of stamina, is illustrative of the care taken to examine and document fish.

For all this, Willughby and Ray had different ideas on precisely how much detail was desirable when compiling a species description. That the former's painstaking descriptions of the plumage colours of birds were met with some apprehension by the latter becomes clear from this passage:

> I must confess that in describing the colours of each single feather he [Willughby] sometimes seems to me to be too scrupulous and particular, partly because Nature doth not in all Individuals, (perhaps not in any two) observe exactly the same spots or strokes, partly because it is very difficult so to word descriptions of this sort as to render them intelligible.[101]

Here, Ray commented on the expedience of tending so much to individual varieties rather than species. He also touched upon the difficulties of putting different shades of colour into words. On this matter, Kusukawa relates a species description in the *Historia piscium* in which Ray considered a certain species of plaice to be of "an unripe olive colour", whereas to Willughby's eyes it seemed more of a brown-greyish colour tending to blue.[102] The discussion of colourisation was a salient one. First of all, because the colours of fish were, indeed, a complicated matter; they might vary according to a fish's age or the season, and usually disappeared once the fish had died.

For Willughby and Ray, therefore, a proper species description was one that referred to one particular species *unambiguously*. Where sixteenth-century naturalists had described species by placing them in a philological context,

99 Since Skippon notes on drawing Mi LM 25/12 that it portrays "the second sight" it is possible that the drawing was originally preceded with another.

100 NUL, Mi LM 25/14; Kusukawa, "*Historia Piscium* (1686) and Its Sources," 316.

101 Willughby, *Ornithology*, preface, sig. A3r. See also: Birkhead, Smith, Doherty and Charmantier, "Willughby's Ornithology," 269–270.

102 Kusukawa, "*Historia Piscium* (1686) and Its Sources," 315.

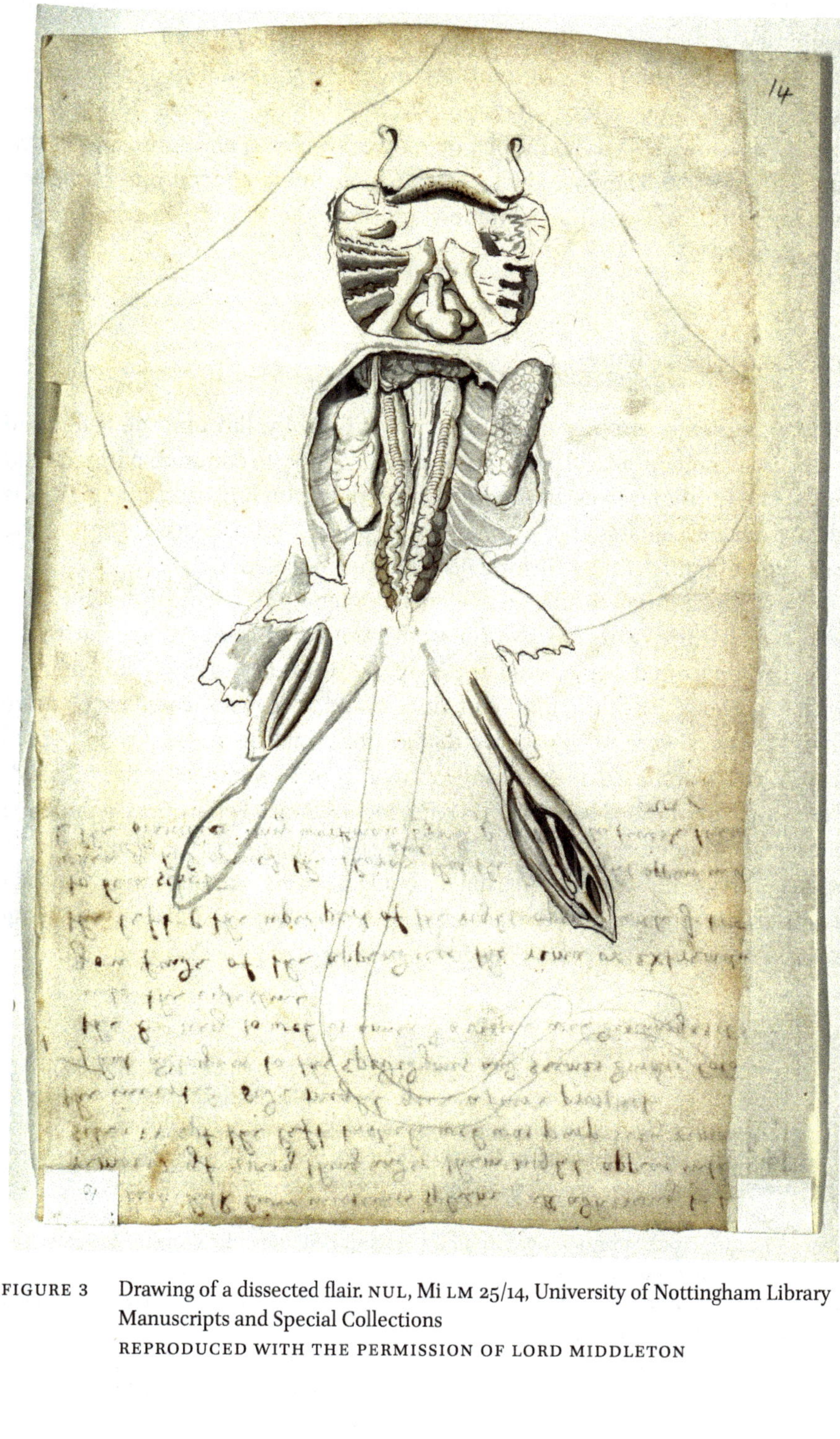

FIGURE 3 Drawing of a dissected flair. NUL, Mi LM 25/14, University of Nottingham Library Manuscripts and Special Collections
REPRODUCED WITH THE PERMISSION OF LORD MIDDLETON

combined with accounts of their physiological aspects, such lengthy, encyclopedic descriptions were less appropriate for Willughby and Ray's purposes. They centered their descriptions around clearly established characteristic marks and only seldom incorporated the cultural context of a species, strictly disavowing interpretations of species of plants and animals as signs and allegories.[103] They sought to establish an unambiguous differentiation between species as well as to understand how these were morphologically interlinked.[104]

4 The Best Figures

Illustrations were another opportunity for Willughby, Ray and the Fellows to emphasise their empirical program. Images, contrary to common parlance, did not speak for themselves, but required interpretation and assessment. This is clear from the debates that erupted while policy and practice regarding the inclusion of figures in the *Historia piscium* were decided upon.

These illustrations display considerable diversity in terms of design. They were culled from a wide range of sources: about two thirds of them were taken from earlier printed works, while the rest were designated as 'new' and marked with a dagger.[105] The designation of 'new' may not have indicated much more than that they were newly-acquired, and this could mean they were copied from newly-purchased manuscripts, drawn from specimens in the Society's Repository, or perhaps had recently been received by a Fellow from a personal contact.[106] Though expensive, Ray probably considered these engravings to be worth every penny. When preparing his book of plants for publication, Ray had asserted that for such a history to lack images would be as remiss as producing an atlas without maps.[107] That images were sources of knowledge on a par with

103 William B. Ashworth has characterised this way of looking at the natural world as the emblematic worldview in his "Natural History and the Emblematic Worldview," in *Reappraisals of the Scientific Revolution*, eds. David C. Lindberg and Robert S. Westman (Cambridge: Cambridge University Press, 1990), 303–332 and "Emblematic Natural History of the Renaissance," in *Cultures of Natural History*, eds. Nicholas Jardine, James Secord and Emma C. Spary (Cambridge: Cambridge University Press, 1996), 17–37.

104 Birkhead, Smith, Doherty and Charmantier, "Willughby's Ornithology," 269.

105 'Figurae Novae, quae non paucae sunt, † notantur'; on how the daggers have not been consistently applied, see Kusukawa, "The *Historia Piscium* (1686)," 186.

106 A detailed overview of the sources for the images can be found in Kusukawa, "*Historia Piscium* and Its Sources," 318–333.

107 Alexander Wragge-Morley, *Aesthetic Science: Representing Nature in the Royal Society of London, 1650–1720* (Chicago: Chicago University Press, 2020), 106; chapter 4 esp. examines the interrelations between verbal and textual depictions and their affects, 106–134.

texts has been pointed out by Kusukawa, who has noted Ray treated "textual description and visual illustration as equally referring to a possibly real fish."[108] For the authors of the *Historia piscium*, illustrations were important sources of knowledge and thus an important component of its output.

When the *Historia piscium* was being prepared for print, the selection of suitable illustrations caused consternation. At the meeting of 18 March 1685 (OS), a letter from Ray to Robinson was read aloud which stated that "with respect to the designs for the cuts, he [Ray] said, that he had several drawn from life, and had made references to the places in authors, where the best figures were extant."[109] This report was not received well. On the very same day the meeting was held, Robert Plot (1640–1696) had penned a response to Francis Aston (1645–1715), who had written him on the matter. During the following meeting, on 25 March 1685 (OS), Plot's response was shared, which suggested that Ray's statement had "much lessened the opinion concerning that history".[110] It had been presumed by the Society that *all* the draughts would be taken "from the life, where as it was now found, that the cuts must be picked up here and there out of books".[111] Now that doubt had been cast on the quality of the illustrations, the intended printer John Fell (1625–1686), who besides being the bishop of Oxford founded the University Press, was hesitant to go ahead with publication.[112] He was not willing to commit to publishing the work "till he had seen what it was; and that therefore those draughts, which were ready, should be sent thither".[113]

An intervention was required. As the Royal Society had decided to finance the work, they set up a committee to oversee the printing process and subject it to quality control. This committee consisted of several Fellows and included Lister, Robinson and Aston, as well as Ray himself.[114] Under the adage that he who pays the piper, calls the tune, the assembling and evaluating of illustrations became a collective project of the Society.[115] When the Fellows found out, for example, that Henry Hunt (d.1713) possessed some of the illustrations of birds and fish made by bishop of Chester John Wilkins (1614–1672), he was "ordered to get the plates of the fishes rolled off against the next meeting, in order that

108 Kusukawa, "The *Historia Piscium* (1686)," 183.
109 Birch, *History of the Royal Society*, vol. 4, 380.
110 Ibid., 382.
111 Ibid.
112 Ibid. See also: Roos, *Web of Nature*, 321.
113 Birch, *History of the Royal Society* vol. 4, 382.
114 Ibid.
115 Ibid., 380; Sachiko Kusukawa, "Picturing Knowledge in the Early Royal Society: The Examples of Richard Waller and Henry Hunt," *Notes and Records of the Royal Society of London* 65, no. 3 (2011): 280.

the Society might judge, whether they would be useful to this book."[116] The debate on suitable drawings for the *Historia piscium* suggests there was no consensus among the Royal Society members with regard to what constituted an epistemologically sound image. Furthermore, these kinds of back and forths between the Fellows not only signal the weight attached to selecting proper illustrations, but also invite reflection on what options they had.

Illustrations played a considerable yet not necessarily clear-cut role in the early Royal Society. Study of the Fellows' visual practices shows that they valued images for communicating, claiming, and proving observations, ideas and theories; and that they took care to copy and preserve illustrations in their own archives, although they did not include every single drawing or graph.[117] In certain instances, they held animus against images; the aforementioned Hooke, who authored a sumptuously illustrated book on microscopic entities, approached images with some wariness as he felt that they could sway the passions and cloud the intellect.[118] All in all, there was not one clear, overarching epistemological programme for images on the part of the Society, and one ought not ascribe too much coherence to the Fellows' approaches towards illustrations.[119]

As the Fellows had feared, a fair share of the illustrations in the *Historia piscium* were copied from the books of Aldrovandi, Belon, Gessner, Rondelet and Salviani. The copying of images from other books was itself a well-established practice. Although it can be difficult to reconstruct such fililations with absolute certainty, it is clear that early modern authors generally drew upon each other's illustrations, just as they drew upon each other's texts.[120] Take for example the image of the sea serpent in Belon's book on aquatic animals (1553) that depicts it coiled as if it were a rope. A very similar image of the same species can be found in Salviani's work published a year later; while it is not an exact copy (especially in the design of the head), it nonetheless corresponds to it in specific details. Comparably coiled, it has the same curl in the tail and an open mouth showing its teeth. A few years later, Gessner's 1558 book displays

116 Birch, *History of the Royal Society* vol. 4, 380.

117 See: Sietske Fransen, Katherine M. Reinhart and Sachiko Kusukawa, "Copying Images in the Archives of the Early Royal Society," *Word & Image* 35, no. 3 (2019): 256–276.

118 Henderson, "Robert Hooke and the Visual World of the Early Royal Society," 421.

119 Sachiko Kusukawa, "The Early Royal Society and Visual Culture," *Perspectives on Science* 27, no. 3 (2019): 381.

120 Kusukawa, *Picturing the Book of Nature*, 64–69. A careful study on the reuse of botanical woodblocks can be found in Jessie Wei-Hsuan Chen, "A Woodblock's Career: Transferring Visual Botanical Knowledge in the Early Modern Low Countries," *Nuncius* 35, no. 1 (2020): 20–63.

an image unmistakeably similar to that of Salviani, albeit with some slight variations in the teeth. Lastly, the depiction of the sea serpent in the *Historia piscium* shows clear resemblance to Salviani and is captioned with '*Serpens marinus Salviani*', which, as said, was also similar to the serpent depicted in Belon.[121] This cursory comparison shows that natural historical images were quite freely copied and adapted.[122]

With the exception of Salviani, the abovementioned authors used woodcuts, the standard medium in the sixteenth century for the production of images. Salviani deployed engravings, a form of intaglio printing in which lines were incised on metal plates (often of copper).[123] While this technique had been in use since the fifteenth century, it was not always the preferred option due to its high cost and because it meant that illustrations had to be printed separately from the text.[124] The advantage of intaglio printing was resolution, however, and while some artists had the skill to work very fine detail into woodcuts, copper engravings generally lent themselves to more precise depiction. This is perhaps why the Fellows preferred them: they used almost three quarters of Salviani's engravings.[125] For the *Historia piscium*, all the illustrations, either copied or designed afresh, were engraved on plates. In this, the printer Fell advised that only one hand should be employed to ensure stylistic consistency.[126] This advice was not heeded; it appears that no less than eighteen engravers were commissioned.[127] While we now know the names of a considerable number of these engravers, this is not the case for the artists who made the initial drawings.

Although Willughby and Ray were careful to put the exact colour of birds and fish in writing, the engravings in the *Ornithologia* and the *Historia piscium* appeared without colour.[128] It was, at this time, possible but not necessarily common to have copies of natural historical works coloured in the workshop of a printer, where colourists often worked from a master copy. This could,

---

121 Cf. Salviani, *Aquatilium animalium historiae*, tab P1 and *Hist. pisc.*, tab G4.

122 See also: Angela Fischel, *Natur im Bild: Zeichnung und Naturerkenntnis bei Conrad Gessner und Ulisse Aldrovandi* (Berlin: Gebr. Mann, 2009), 103–106.

123 Kusukawa, *Picturing the Book of Nature*, 32–33.

124 Ibid. 32–34.

125 Kusukawa, "*Historia Piscium* (1686) and Its Sources," 307.

126 Roos, *Web of Nature*, 323.

127 The individual plates do not specify who worked on which engraving. Most of the engravers are identified by name in Kusukawa, "The *Historia Piscium* (1686)," 191.

128 A (partially) coloured copy of the *Ornithology* is described, however, in Robert Montgomerie and Tim R. Birkhead, "Samuel Pepys's Hand-Coloured Copy of John Ray's 'The Ornithology of Francis Willughby' (1678)," *Journal of Ornithology* 150, no. 4 (2009): 884.

however, be tricky; Gessner, for example, pointed out that the colourists employed by the publisher for his *Historia animalium* had carried out their task all too carelessly.[129] The colourisation of engravings, which might be done either before or after purchase, added a considerable additional expense on the part of the buyer.

When Ray announced that some of the drawings for the book were taken from life, Plot contended that all of them should have been. The multivalent meaning of the phrase 'from the life' (and its Latin cognate *ad vivum* as well as counterparts in other languages) has been amply discussed in the last years by art historians and historians of science alike.[130] Kusukawa has noted that one should be careful to attach too much coherence to the ways in which this term was used. It did not necessarily imply that the artists had any direct experience with the object they portrayed.[131] In fact, the qualification of '*ad vivum*' by no means confirmed the existence of the thing that was depicted.[132] It could mean, rather, that a certain object was painted as lifelike enough to evoke in the spectator the idea that the thing was really there.[133] It could also serve as in indication of an intention on behalf of the artist to have rendered the animal or plant as accurately as possible.[134] None of the plates in the *Historia piscium* carry an inscription stating that they were done 'from the life'; we have seen, however, that Ray did distinguish between images that were done 'from the life' and those that had been copied from books.

The ambiguity of the qualification 'from the life' can be illustrated by a species of pufferfish described and depicted in *Historia piscium*. It concerns the *Guamaica atinga*, as it was called by the Tupi people of the coastal regions of Brazil. Neither Willughby nor Ray had ever seen the species alive. They had taken the species description from the *Historia naturalis Brasiliae* (Amsterdam, 1648), the natural historical work of Georg Marcgraf (1610–1644) and Willem Piso (1611–1678).[135] These authors had travelled through the north-eastern prov-

---

129 Kusukawa, *Picturing the Book of Nature*, 76.

130 The historiography on the theme has been summarised and analysed by Thomas Balfe and Joanna Woodall, "Introduction: The Lives of *Ad vivum*," in *Ad vivum? Visual Materials and the Vocabulary of Life-Likeness in Europe before 1800*, eds. Thomas Balfe, Joanna Woodall and Claus Zittel (Leiden: Brill, 2019).

131 Sachiko Kusukawa, "*Ad vivum* Images and Knowledge of Nature in Early Modern Europe," in Balfe, Woodall and Zittel, *Ad vivum?*, 112.

132 Ibid., 90–91.

133 Ibid.

134 Egmond, *Eye for Detail*, 94.

135 See also: Peter Whitehead, "George Markgraf and Brazilian Zoology," in *Johan Maurits van Nassau-Siegen: A Humanist Prince in Europe and Brazil*, ed. E. van den Boogaard (The Hague: Mauritshuis, 1979), 424–471; *A Portrait of Dutch 17th Century Brasil: Animals, Plants*

inces of Brazil in the service of Johan Maurits van Nassau-Siegen (1604–1679), who had been made the governor-general of this Dutch colony in 1636.[136] Many woodcuts from their work were used in the *Historia piscium*. In the case of the *Guamaica atinga*, however, rather than simply copying the published woodcut that was based on an *in situ* drawing in Brazil, a drawing was designed afresh.

This drawing (Figure 4), which is among the Willughby legacy, was one of those made from a specimen in the Royal Society's Repository.[137] It was one among the six specimens in the Repository that were drawn and engraved for the *Historia piscium*.[138] The pufferfish was drawn from the life, in the sense that the artist based the drawing from direct access to the specimen at hand, rather than copying an earlier illustration. By including the shadow cast by the pufferfish onto the paper, the artist directs attention to both his own physical presence and that of the specimen. The image thus employs a 'rhetoric of the real.'[139] Obviously, the specimen was no longer alive at this point. The illustrations in the *Historia naturalis Brasiliae* were explicitly declared *ad vivum* in the dedication of the work, though what this entails is not made explicit.[140]

In describing species, characteristic marks were key to Willughby and Ray, and this emphasis manifested itself in the book's illustrations. First of all, most of the depicted fish species are shown from a side elevation, to borrow an architectural term – with the obvious exception of rays and flatfishes, where often both the upper and lower surfaces are shown. This angle of depiction best conveys the morphological characters of fish pertinent for identification and categorisation, such as their overall shape and their fins. Furthermore, they are depicted according to what Janice Neri has dubbed 'specimen logic', where objects are taken from their context and placed within the blank space

---

and People by the Artists of Johan Maurits of Nassau, eds. Peter Whitehead and Marinus Boeseman (Amsterdam: North-Holland Publishing Company, 1989).

136 Recent research has brought to light Johan Maurits's role in the transatlantic slave trade. Carolina Monteiro and Eric Odegard, "Slavery at the Court of the 'Humanist Prince' Reexamining Johan Maurits van Nassau-Siegen and his Role in Slavery, Slave Trade and Slave-smuggling in Dutch Brazil," *Journal of Early American History* 10, no. 1 (2020): 3–32.

137 Drawing of a pufferfish, NUL, Mi LM 25/21.

138 The Middleton collection holds six drawings that served as models for these engravings, see: Kusukawa, "*Historia Piscium* (1686) and Its Sources," 326–327.

139 Cf. *Hist. pisc.* tab I8. For the notion "rhetoric of the real", see Martin Kemp, "Style and Non-Style in Anatomical Illustration: From Renaissance Humanism to Henry Gray," *Journal of Anatomy* 216, no. 2 (2010): 192–208.

140 Paul J. Smith, "Marcgraf's Fish in the *Historia Naturalis Brasiliae* and the Rhetorics of Autoptic Testimony," in Mariana C. Françozo (ed.), *Toward an intercultural natural history of Brazil. The* Historia Naturalis Brasiliae *reconsidered* (New York: Routledge, 2023), 131.

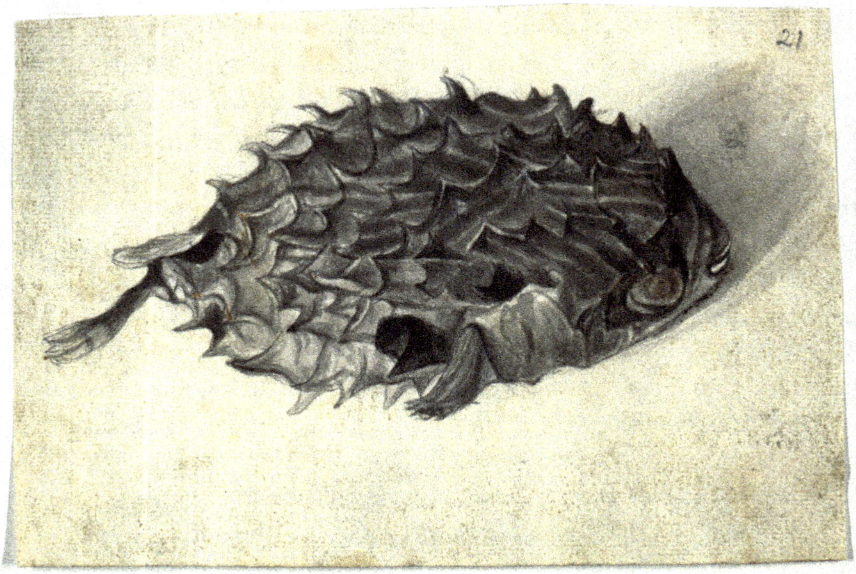

FIGURE 4 Drawing of a pufferfish specimen from Royal Society Repository. NUL, Mi LM 25/21, University of Nottingham Library Manuscripts and Special Collections

of a page.[141] Both showing fishes from the side and placing them against a white background are common depictive strategies in natural historical works from the sixteenth through to the nineteenth centuries. That did not mean, however, that all characteristic marks were necessarily rendered visible. In the earlier discussion of the differences between the tope shark and smooth hound shark, it was shown how these related to the colour of the irises, their teeth, and their size. None of these features can be inferred from the respective illustrations in the *Historia piscium*, for none of these depictions convey the colour, inner anatomies, or size of their depicted species (Figure 5a/b).[142] The illustrations were thus not in themselves always sufficient to enable species to be told apart.

141 Janice Neri, *The Insect and the Image: Visualizing Nature in Early Modern Europe 1500–1700* (Minneapolis: University of Minnesota Press, 2011), xii; a critique of this term is found in Egmond, *Eye for Detail*, 104.

142 Cf. *Hist. pisc.*, tab B5 and B6.

FIGURE 5A
Engraving of a smooth hound shark. Francis Willughby and John Ray, *Historia piscium* (Oxford: Sheldonian Theatre, 1686), tab B5
UTRECHT UNIVERSITY LIBRARY

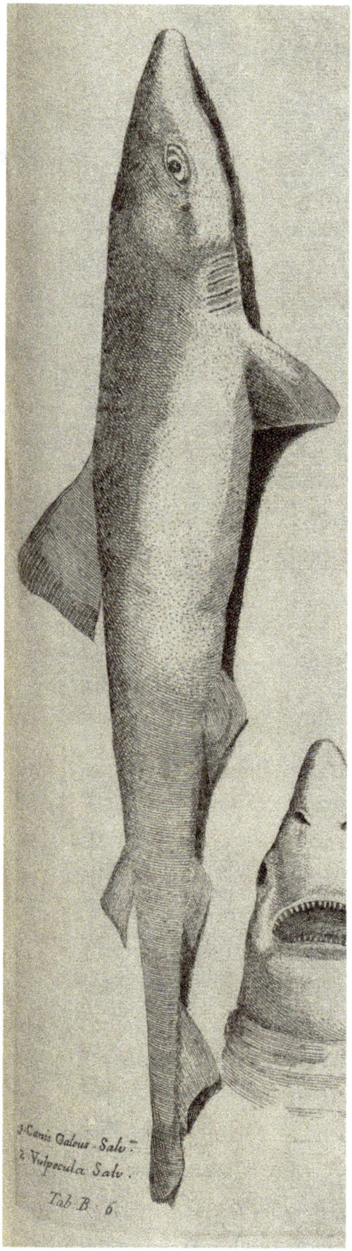

FIGURE 5B
Engraving of a tope shark. Francis Willughby and John Ray, *Historia piscium* (Oxford: Sheldonian Theatre, 1686), tab B6
UTRECHT UNIVERSITY LIBRARY

For Ray, images were nonetheless important because they conveyed knowledge with ease and pleasure.[143] He was certainly pleased with how the illustrations in the *Historia piscium* turned out, as he praised their veracity.[144] At the same time, representations were mediations, as the compilers of the *Historia piscium* were also aware. Authors did not exercise complete control over the process of illustration.[145] It involved many people, from draftsmen to block-cutters or engravers, and printers and colourists. The drawings for the *Historia piscium* of fishes were selected on a case-by-case basis. In this selection process, we recognise the decisions made by sixteenth-century naturalists who had drawings made from life, but who also used images that were sent to them, or copied figures from elsewhere. In fact, a share of these images were copied into the *Historia piscium*. Images that were engraved anew often incorporated new pictorial conventions such as the inclusion of shadows to suggest proximity to the object at hand. The question of what were the 'best figures', in Ray's words, thus had pragmatic answers that were a function of a plethora of factors such as the qualities of the artists, depiction strategies and conventions, and technological possibilities.

5 Conclusion

In the *Historia piscium*, Willughby and Ray defined their subject matter differently than authors had done before them: both with regard to what a fish was (i.e., an animal with fins rather than any creature that dwelled in the water), and as to how species should be described and categorised. Natural history of the sixteenth century was characterised by an encyclopaedic approach to studying nature, one bent upon knowing and describing the particular properties of species of plants and animals based on a wide range of sources, from cookery books and literary works to medicinal treatises. Willughby and Ray, in seeking to reform the study of nature, explicitly bade farewell to this humanistic research tradition, propagating instead a strict focus on describing morphological features of fish through direct observation.

And yet, the book retained a compilatory character, as Willughby, Ray and the Fellows combed through material from a wide range of sources, from earlier books and travel accounts, to loose drawings, and observations communicated in letters. The debates surrounding species descriptions (Willughby and Ray)

143 Wragge-Morley, *Aesthetic Science*, 107.

144 Kusukawa, "The *Historia Piscium* (1686)," 186.

145 Kusukawa, "Illustrating Nature," 97.

and illustrations (Ray and the rest of the Fellows) demonstrate how idealistic choices also had to be pragmatic. Willughby, Ray and the other Fellows were very much reliant on the observations and experiences of others. Studying nature entailed different degrees of observation, as one could see the specimen itself, either dead or alive, or peruse an illustration, read a description, or hear it reported from someone. Fellows thus had to continually evaluate which observations were worthwhile and trustworthy, and which were not. The following chapter will expand on this idea and foreground questions of reliability and credibility. It centres on the ideals of empirical observation as espoused by the Royal Society in general and the *Historia piscium* more specifically, and looks at how these related to the experiences of fishermen and fishmongers. It takes us to the shops, the ports and the fish markets.

CHAPTER 2

Fresh Fish: Observation Up Close in Francis Willughby and John Ray's *Historia piscium*

Fishes were part and parcel of daily life in early modern England.[*] This becomes clear when perusing some of the species descriptions in Willughby and Ray's *Historia piscium*. An annotated copy in the archives of the Royal Society further accentuates this. It is the Society's original, very own copy, and both Tancred Robinson and, later, Cromwell Mortimer (1693–1752) took the liberty of adding their own remarks and observations in the margins of certain species descriptions.[1] As such, it offers insight into the questions that Fellows continued to explore even after the history of fishes was published. A considerable proportion of these notes is dedicated to specifying where in London one might chance upon which species of fish. They reveal, for example, that lampreys could be seen shining in the water of the Thames before fishermen hauled them up in wicker nets, whereas London shops displayed a selection of dab.[2] A dolphin – at that time still considered a fish – taken "in our Channell; very smooth like polisht marble a long snout with 2 rows of teeth on each side, very little Eyes & c. about 4 feet long" could be encountered "at the Ship Tavern at Butcher Row's end near Temple Bar."[3] The swim bladder of the cod counted as a "very luxurious" dish in the city.[4] Any strange fishes caught in the Thames, furthermore, were brought to the Lord Mayor's home.[5] Despite their ubiquitous presence, however, fish were also somewhat elusive. Some of these 'slippery denizens'[6] of the water were difficult to capture, and once caught they

* An adaptation of this chapter has appeared as article: Didi van Trijp, "Fresh Fish: Observation up Close in Late Seventeenth-Century England," *Notes and Records of the Royal Society of London* 75, no. 3 (2021): 311–332.

1 Willughby and Ray, *Historia piscium*, RCN 18574, Library and Archives of the Royal Society (hereafter RS), London. The annotations are the remarks of Tancred Robinson (TR) as inscribed by Francis Aston; later annotations are Cromwell Mortimer's (CM). The copy is also mentioned in Kusukawa, "*Historia Piscium* (1686) and Its Sources," 328.

2 Willughby and Ray, *Historia piscium*, RS, RCN 18574RCN, 96–97, 105 (TR).

3 Ibid., 28 (CM).

4 Ibid., 166 (CM).

5 Birch, *The History of the Royal Society*, vol. 4, 42.

6 The phrase "slippery denizens" comes from Matthew C. Hunter, *Wicked Intelligence: Visual Art and the Science of Experiment in Restoration London* (Chicago: University of Chicago Press, 2013), 69.

promptly began to falter and spoil.[7] Where and how, then, could one establish solid knowledge about these unstable objects of inquiry?

The previous chapter has explained that the aim of the *Historia piscium* was to provide accurate accounts of all fish hitherto known, and to do so in an orderly manner. It has elaborated on how Willughby and Ray focussed on the physical characteristics that fish displayed, and which they had, ideally, selected after close observation of the species at hand. We saw how the materials on which they could draw were rich, and that these encompassed earlier natural historical works, travel accounts, objects in cabinets of curiosities, drawings bound together in books, loose drawings, and observations shared in letters. We also saw how the process of creating precise species descriptions and selecting suitable accompanying illustrations every now and then provoked discussions regarding what ought to be the proper selection criteria. This chapter analyses such matters of evaluation in more detail. It looks closely at how the observation of fish took place in practice, and at how observations were assessed as reliable and credible and thence incorporated into the *Historia piscium*.

In addressing these matters, the engraved title page (Figure 6) made by the Dutch painter and printmaker Paul van Somer II (1649–1714) entreats the reader to take a closer look. Set against the backdrop of an Arcadian fishing port, several people tend to the arrival of fresh fish announced by a herald blowing a large conch shell.[8] Fishermen in loincloths haul in their nets. Two men dressed in tunics examine the scene, one of whom gestures at the catch. Just below them, a female figure in a helmet, possibly a reference to Minerva, the Roman goddess of wisdom and the arts, draws the specimen that is set before her. A garland of fish lines the sides and top of the frontispiece; the pufferfish, turbot, and hound shark are copied from the engraved plates of the book.[9] These depictions are decidedly different from the dolphin, taken from classical iconography, which adorns the lower left corner of the engraving. The colossal fish in the foreground, containing the book's imprint and its

7 This elusiveness is explored in Elspeth Graham, "Ways of Being, Ways of Knowing: Fish, Fishing, and Forms of Identity in Seventeenth-Century English Culture," in *Animals and Early Modern Identity*, ed. Pia F. Cuneo (Farnham: Ashgate, 2014).

8 Anna Marie Roos has suggested that the ship on the title page is a visual nod to the one featured on the engraved title page to Bacon's *Instauratio magna* (London: John Bill, 1620). See Roos, *Web of Nature*, 325.

9 For a discussion of title pages of natural historical works on fish, see: Paul J. Smith and Didi van Trijp, "Dynamiques européennes de l'humanisme érudit dans l'histoire naturelle. Le cas de l'ichtyologie, de Belon, Rondelet et Gessner à Willughby et Ray," in *L'humanisme à l'épreuve de l'Europe* (*XV*^e^–*XVI*^e^ *siècles*), eds. Denis Crouzet, Elisabeth Crouzet-Pavan, Philippe Desan and Clémence Revest (Ceyzérieu: Champ Vallon, 2019), 167–181.

FIGURE 6 Engraved title page of *Icthyographia* (1685), Paul van Somer II. RS.9493, The Royal Society

affiliation to the Royal Society in its gaping mouth, is rendered in a similarly stylized manner.[10] To the right of this creature, a female figure reposes on a jug from which water is pouring, adding to the sense of flow and movement of the scene. All in all, the title page conjures an image of exuberance and abundance. Considering that frontispieces of early modern works of natural history and philosophy often present a visual programme of a book's contents, this one brings together the various sources available for finding knowledge about fish: classical accounts, illustration, and first-hand observation.[11]

One source of knowledge about fish is displayed particularly prominently on the title page: namely, those people practically engaged with fish, such as fishermen and fishmongers. The nature and extent of the contributions of these practical men can be inferred both from the *Historia piscium* itself, and from other source materials related to the book and its authors, such as natural historical manuscripts, minutes of Royal Society meetings, and letters to and from the Fellows.[12] While recent studies of the *Historia piscium* do mention their contributions, they do so only in passing and a thorough analysis has not as yet been undertaken.[13] Examining the interactions between fishermen and Fellows offers key insight into what was considered valuable knowledge about fish, and especially who could be counted on to produce said knowledge.

The first part of this chapter embeds the *Historia piscium* in the broader social-cultural context of knowledge production particular to the Royal Society, which valued the (direct) experiences of trustworthy observers. It discusses how we can position fishermen and fishmongers in the Society's circle of informants. The second part examines why Fellows turned to fishermen, and argues that, in natural historical studies, a supply of fresh fish was often much preferred to examining preserved specimens or illustrations. The third

10 A print proof of the engraved title page, in which both title and affiliation have yet to be inserted, can be found in NUL, Mi LM 24/170.

11 Volker R. Remmert, "'Docet parva picture, quod multae scripturae non dicunt.' Frontispieces, Their Functions, and Their Audiences in Seventeenth-Century Mathematical Sciences," in *Transmitting Knowledge: Words, Images, and Instruments in Early Modern Europe*, eds. Sachiko Kusukawa and Ian Maclean (Oxford: Oxford University Press, 2006), 240; see also idem, *Picturing the Scientific Revolution* (Philadelphia: Saint Joseph's University Press, 2011).

12 It was indeed often men; no fishwives figure in the sources examined here. In England, fishwives were not allowed to sell inside public marketplaces, see: Alena Buis, Christi Spain-Savage and Myra E. Wright, "Attending to Fishwives: Views from Seventeenth-Century London and Amsterdam," in *Mapping Gendered Routes and Spaces in the Early Modern World*, ed. Merry E. Wiesner-Hanks (Farnham: Ashgate, 2015), 193.

13 Other studies of the *Historia piscium* can be found in Kusukawa, "The *Historia Piscium* (1686),"; idem, "*Historia Piscium* (1686) and Its Sources," and Raven, *John Ray*, 339–370.

part addresses how these practical men contributed to the identification of, and distinction between, particular species and remarked on specific behaviour. As such, it looks into what fishermen knew about fish, and the extent to which Willughby, Ray and the other Fellows considered them as useful and reliable sources. As we will see, experience was an essential factor in evaluating these men's claims and observations. The emphasis that the Society placed on direct observation as necessary in the establishment of accurate accounts of species required both a wide range of observers and an assessment of these observers on the part of the Fellows.

1 A Wider Cast

The variety of sources displayed on the title page of the *Historia piscium* is also reflected in the text itself, such as in discussion of the peculiar way in which the salmon every so often leaps out of the water:

> The *salmon* constantly presses forward against the stream, and when it encounters in its ascent an enclosure or another obstacle of this kind, it seizes, after it has bent its body in a circle, its tail with its mouth, and, while it holds fast to this [i.e., its tail], it, releasing [its grip] again, with great force, leaps across it. Author of *De natura rerum* with *Gesner*. We have heard multiple times of many fishermen that this happens continually. That salmon are most agile in jumping we confirm willingly, and our daily experience confirms this: but what is told about the seizing of the tail seems to us less plausible.[14]

Several layers of observation come together in this passage. It begins with a medieval account of this phenomenon, by Thomas of Cantimpré (1201–1272), as cited in Gessner.[15] While this is illustrative of the extent to which Willughby

14 '*Salmo* adversus fluvios perpetuo nititur, cumque in ascensu sepem vel aliud hujusmodi obstaculum invenerit, in circulum flexo corpore caudam ore apprehendit, eamque mordicus tenens, iterumque dimittens magno impetu transilit. *Author de natura rerum* apud *Gesner*. Hoc à plurimis piscatoribus assidue fieri multoties audivimus. Quod Salmones ad saliendum agillimi sunt, libenter concedimus, & experientia quotidiana confirmat: verum quod de caudae apprehensione fertur minus verisimile nobis videtur.' *Hist. pisc.*, 191–192.

15 Cf. Gessner, *Hist. anim.* IIII, 974 and Thomas of Cantimpré, *De natura rerum*, ed. Helmut Boese (Berlin: De Gruyter, 1973), 270. See also: Baudouin Van den Abeele, "Conrad Gessner als Leser mittelalterlicher Enzyklopädien," in Leu and Opitz, *Conrad Gessner*, 15–28.

and Ray drew on the works of earlier Renaissance authors, as has been argued in the previous chapter, they also did not take such accounts at face value. Willughby and Ray verified this account, not once, but in multiple instances, and not with one, but with many fishermen – who, furthermore, confirmed that they saw this happening all the time. This, in itself, however, still did not settle the matter of the salmon's strange behaviour. While Willughby and Ray's own, daily experiences confirmed the tenor of the report, namely that salmon are nimble jumpers, they remained sceptical about its specifics, namely the manner in which the salmon gripped and released its tail, which they had not seen themselves. The *Historia piscium* contains more passages like these, which cite observations from past and/or present sources before concluding with the authors' own verdict on the matter.[16]

The previous chapter has addressed how the publication of the book was the result of a collective effort of the Fellows of the Royal Society. They were closely involved in selecting what merited inclusion in the work. As we will see, the work can also be recognised as a product of the Society in its insistence on knowledge derived from direct experience with the object of study. This certainly did not mean that the Fellows no longer consulted texts, but that these texts were not considered sufficient in themselves.[17] In the epilogue to the *Historia piscium*, Ray contended that it would "bring across exactly these things which were either observed by ourselves and our friends, or which had proper witnesses and authors, worthy of our trust."[18] While earlier authors counted as credible past witnesses, their written observations would, ideally, be corroborated with those of contemporary ones. Indications of direct observation are present in the fish book in various ways. Willughby and Ray, for example, added 'I have seen' [*vidi*] or 'we have seen' [*vidimus*] to certain species descriptions – this variation of the singular and plural form being another indication of the complicated layers of authorship discussed in Chapter 1.[19] These kinds of pithy phrases, specifying whether one had acquired knowledge of a thing with one's own eyes or from hearsay, had already been proposed by Bacon in order to indicate the reliability of a statement.[20]

---

16 See also, for example, *Hist. pisc.*, 105, 229 and 342.

17 As discussed in Fabian Krämer, *Ein Zentaur in London: Lektüre und Beobachtung in der frühneuzeitlichen Naturforschung* (Korb: Didymos, 2014).

18 '[…] duntaxat tradere quae vel à nobismetipsis & amicis observata essent, vel idoneos & fide dignos testes & auctores haberent.' *Hist. pisc. app.*, sig. Hv.

19 For example, *vidi* usually (but not always) refers to Ray, and *vidimus* usually to both Willughby and Ray, but sometimes only to Ray.

20 See: Bacon, *Parasceve ad historiam naturalem*, OFB XI, 467 and Andrew Peter Langman, "'Beyond, both the Old World, and the New': Authority and Knowledge in the Works

In other cases, Willughby and Ray punctuated statements with appeals to 'experience' [*experientia*], as in the case of the salmon. The exact meaning of this term was far from fixed in the early modern period.[21] Although Peter Dear argued that, in the early years of the Royal Society, 'experience' was used for witnessing or participating in a particular, singular event tied to a specific moment, rather than for generalised statements on universal phenomena (in the Aristotelian sense of the term),[22] it seems that the term figures in both senses within the *Historia piscium*. The usage of the term ranges from the more general 'experience agrees' [*experientia constat*], to the collective 'experience has taught us' [*experientia didicimus*] to the more specific, individual 'that which my experience has confirmed' [*id quod experientia mihi confirmavit*].[23]

In the previous chapter, it has been explained that emphasis on first-hand observation (for which the terms *observatio* and *autopsia* gained currency) rose steadily from the early sixteenth century onwards.[24] It has also been remarked that Society's foregrounding of direct experience as the foundation of natural knowledge owes much to Bacon's work.[25] Experience of nature might be gained, Bacon had stated, through hunting, husbandry, gardening, shepherding, animal breeding and travelling, among other things.[26] "The materials for the intellect", he wrote, "are so widely spread out that they ought to be sought out and gathered in (as if by agents and merchants) from all sides."[27] A similar sentiment can be read from in the words of Sprat, when he wrote that

---

of Francis Bacon, with Special Reference to the *New Atlantis*" (PhD. diss., Queen Mary University of London, 2007), 243–244.

21 See for example, Alberto Vanzo, ed., *Experience in Natural Philosophy and Medicine*, special issue of *Perspectives on Science* 24, no. 3 (2016): 255–379; Peter Dear, "The Meanings of Experience," in Daston and Park, *Cambridge History of Science*, vol. 3, 106–131; and Ogilvie, *Science of Describing*, 17–23.

22 Peter Dear, *Discipline and Experience: The Mathematical Way in the Scientific Revolution* (Chicago: Chicago University Press, 1995), 13–14.

23 *Hist. pisc.*, 7, 9 and 246.

24 Gianna Pomata, "Observation Rising: Birth of an Epistemic Genre, ca. 1500–1650," in *Histories of Scientific Observation*, eds. Lorraine Daston and Elizabeth Lunbeck (Chicago: University of Chicago Press, 2011), 45–80.

25 As it is impossible to do justice to the intricacies of Bacon's epistemology or historiography here, I mention particularly on natural history: *Francis Bacon and the Reconfiguration of Early Modern Natural History*, eds. Guido Giglioni, Dana Jalobeanu and Sorana Corneanu, special issue of *Early Science and Medicine* 17, no. 1/2 (2012): 1–271.

26 Guido Giglioni, "Learning to Read Nature: Francis Bacon's Notion of Experiential Literacy (*Experientia Literata*)," *Early Science and Medicine* 18, no. 4/5 (2013): 409.

27 Francis Bacon, *The Instauratio Magna Part II: Novum Organum and Associated Texts*, eds. Graham Rees and Maria Wakely (Oxford: Clarendon Press, 2004), 451. Henceforth abbreviated as OFB XI.

knowledge was to be gathered "from the Shops of Mechanicks; from the Voyages of Merchants; from the Ploughs of Husbandmen; from the Sports, the *Fishponds*, the Parks, the Gardens of Gentlemen".[28]

Bacon, however, also held that one would be "forever tossed and turned on the waves of experience" if it were pursued without clear course.[29] Those who wished to interpret nature required a degree of 'literate experience' [*experientia literata*].[30] Characteristically perhaps for Bacon's at times somewhat opaque manner of formulation, historians of science have grappled with what exactly this kind of experience signified, and offered various interpretations of this notion.[31] Bacon himself stated that "no discovery should be sanctioned save that it be put in writing. Only when that becomes standard practice, with experience at last becoming literate, should we hope for better things."[32] Sophie Weeks has highlighted how the primary difference between literate and illiterate experience was not a matter of erudition, but rather one of mediated access to nature. This meditation entailed a disciplined examination of nature, in which it was "set down and presented in suitable order" rather than investigated in all its fecundity at random.[33] Whether this then meant that such an orderly way of probing nature was restricted to the educated or learned is another question. Deborah Harkness has contended that Bacon's precepts for obtaining true and certain natural knowledge harked back to the daily vernacular science that was practiced in the streets of Elizabethan London.[34]

Fishermen and fishmongers, as attentive observers of nature, were consulted broadly throughout both the classical and early modern period. In their study of nature, Aristotle and Pliny the Elder drew on the reports of those whose

28 Sprat, *History of the Royal-Society*, 72; emphasis mine.

29 Bacon, *Novum organum*, OFB XI, 16–17.

30 A notion that Bacon developed in ibid., 158–159, according to Cesare Pastorino, "Weighing Experience: Experimental Histories and Francis Bacon's Quantitative Program," *Early Science and Medicine* 16, no. 6 (2011): 543.

31 Cf. Dana Jalobeanu, "Disciplining Experience: Francis Bacon's Experimental Series and the Art of Experimenting," *Perspectives on Science* 24, no. 3 (2016): 324–342; Giglioni, "Learning to Read Nature," 405–434; Sophie Weeks, "The Role of Mechanics in Bacon's *Great Instauration*," in *Philosophies of Technology: Francis Bacon and his Contemporaries*, eds. Claus Zittel, Gisela Engel, Romano Nanni and Nicole C. Karafyllis (Leiden: Brill, 2008), 133–195; Lisa Jardine, "*Experientia literata* or *Novum organum*? The Dilemma of Bacon's Scientific Method," in *Francis Bacon's Legacy of Texts: 'The Art of Discovery Grows with Discovery'*, ed. William A. Sessions (New York: Ams Press, 1990), 47–67.

32 Bacon, *Novum organum*, OFB XI, 159.

33 Bacon, *Novum organum*, OFB XI, 215 as cited in Weeks, "Mechanics in Bacon's Great Instauration," 172.

34 Harkness, *The Jewel House*, 213.

experience of nature stemmed from practice, such as fishermen, huntsmen, shepherds, farmers and seafarers.[35] These communities were important for the sixteenth-century humanists who took a philological approach to studying nature, incorporating the common names of species alongside their Greek and Latin names.[36] Naturalists like Rondelet and Belon, for example, conversed with fishermen on their observations of Mediterranean marine life in addition to perusing learned books; a practice that Florike Egmond has referred to as 'fieldwork once removed.'[37] Gessner, too, stated that he benefited from the knowledge of fishermen, and attributed a higher value to direct-hand observation than he did to natural knowledge of the textual kind.[38] Aldrovandi's correspondents wrote him about their trips to fish markets to glean information about species from fishmongers.[39] Monica Azzolini has shown how, in early seventeenth-century Rome, naturalists like Johannes Faber (1574–1629) made ample use of a plurality of oral sources including fishermen, merchants, and servants, when investigating beached whales.[40]

As we will see, these interactions take on a new meaning with the emergence of scientific societies in the seventeenth century. Learned societies of this kind emerged over the course of the seventeenth century, in Florence, Rome, Schweinfurt and Paris, amongst other places. As has been argued, most forcefully for the English context, membership of such societies, which was usually restricted to those of the upper classes, was closely linked to matters of trustworthiness.[41] This worked in two directions: if its members thought someone was credible, they selected him to figure in their midst; conversely, belonging to such a group considerably heightened one's credibility. When

35 Maclean, "White Crows, Graying Hair, and Eyelashes: Problems for Natural Historians in the Reception of Aristotelian Logic and Biology from Pomponazzi to Bacon," in Pomata and Siraisi, *Historia*, 157.

36 Pietro Daniel Omodeo, "The Invisible Fisherman: The Economy of Water Knowledge in Early Modern Venice," *Ichthology in Context (1500–1880)*, 366.

37 Florike Egmond, "On Northern Shores: Sixteenth-Century Observations of Fish and Seabirds (North Sea and North Atlantic)," in *Naturalists in the Field: Collecting, Recording and Preserving the Natural World from the Fifteenth to the Twenty-First Century*, ed. Arthur MacGregor (Leiden: Brill, 2018), 131.

38 Anthony Grafton, "Philological and Artisanal Knowledge Making in Renaissance Natural History: A Study in Cultures of Knowledge," *History of the Humanities* 3, no. 1 (2018): 43–45.

39 Findlen, *Possessing Nature*, 176–177.

40 Monica Azzolini, "Talking of Animals: Whales, Ambergris, and the Circulation of Knowledge in Seventeenth-Century Rome," *Renaissance Studies* 31, no 2 (2017): 318.

41 Shapin, *A Social History of Truth*, 122–123; Shapin and Schaffer, *Leviathan and the Air-Pump*, 58.

discussing Faber's report on the whale in the *Historia piscium*, for example, Ray noted with some insistence that the Roman was a member of the Accademia dei Lincei.[42] In the Royal Society, the existing convention of assigning reliability to those of higher social status remained in place when observing and interpreting natural phenomena.[43]

This did not mean, however, that status was the sole criterion of credibility.[44] While those from a genteel background were generally seen as trustworthy, they were also considered prone to bending their observations to fit with preconceived ideas.[45] Philippa Hellawell has argued that credibility was not the exclusive prerogative of one particular social group, but that it could be shared, albeit still attributed in various degrees, among people of various backgrounds.[46] Felicity Henderson has submitted that the Royal Society, as an institution, relied on "the activities and expertise of a wider penumbra of individuals" than that of the Fellows themselves.[47] Certain individuals within the Society itself blurred social boundaries, such as Hooke, who, as son of a curate, required financial support for his studies of nature.[48] Despite being employed as Curator of Experiments, regarded as a lesser position than that of Fellow because of the paid labour involved, he took part in natural philosophical debates and was elected Fellow in 1663.[49]

Experiments held a special place in the deliberations of the early Royal Society. Bacon had contended that experiments served to deliberately seek out a certain experience, as opposed to mere experience which derived from 'accident' – allotting an active role to the observer, rather than a passive one.[50] While the Fellows seem to have had their own approaches to the meaning and

42 Kusukawa, "*Historia Piscium* (1686) and Its Sources," 333.

43 Shapin and Schaffer, *Leviathan and the Air-Pump*, 58.

44 Barbara Shapiro, *A Culture of Fact: England 1550–1720* (Ithaca: Cornell University Press, 2000), 140.

45 Peter Dear, "*Totius in verba*: Rhetoric and Authority in the Early Royal Society," *Isis* 76, no. 2 (1985): 156.

46 Philippa Hellawell, "'The Best and Most Practical Philosophers': Seamen and the Authority of Experience in Early Modern Science," *History of Science* 58, no. 1 (2019): 32.

47 Henderson, "Robert Hooke and the Visual World of the Royal Society," 397.

48 Steven Pumfrey, "Ideas above His Station: A Social Study of Hooke's Curatorship of Experiments," *History of Science* 29, no. 1 (1991): 4.

49 Steven Pumfrey, "Who did the Work? Experimental Philosophers and Public Demonstrators in Augustan England," *British Journal for the History of Science*, 28, no. 2 (1995): 153.

50 Bacon, *Novum organum*, 131. It is important to note that he did not apply the notions *experientia, experimentum*, and *observatio* particularly strictly, see Lorraine Daston, "The Empire of Observation, 1600–1800," in Daston and Lunbeck, *Histories of Scientific Observation*, 83. 81–113.

use of experiments, it is clear that several of them took to performing them as a way of understanding nature's intriguing properties.[51] Regarding fish, they pondered such questions as: did they breathe? How did these creatures move in the water? How did they spawn, and how long could they go without food? The minutes of meetings found in the Journal Books of the early 1660s reveal that the Society's Operator, whose task it was to facilitate experiments and make inquiries, was ordered several times to collect and keep fish for experiments.[52] He was also instructed to ask fishermen how long they could keep their fish alive without feeding them.[53] Furthermore, the minutes indicate that "all those [present at the Society], that had the opportun[it]y, were desired to make several Experiments in several fish, concerning their growth."[54]

Although the precise set-up of these experiments is not always disclosed in the minutes, the careful reports published in the *Philosophical Transactions* may give us an idea.[55] Around 1670, Robert Boyle had a gudgeon placed into a 'Pneumatical Engin', or air pump.[56] Of course, Boyle and his company are known to have inserted various small animals into this device, including birds, mice and snakes.[57] The experiment on the gudgeon, "far from being the first" that had been done on a fish with this sort of instrument, was devised to show what happened to a fish when "it should be kept for some hours together from all supply of fresh Air."[58] Although after mostly all of the air was removed "there appeared a great store of Bubbles all about the Fish", no definitive conclusions could be drawn.[59] The specimen lived for some ten days more; Boyle's

51 Peter Anstey, "Philosophy of Experiment in Early Modern England: The Case of Bacon, Boyle and Hooke," *Early Science and Medicine* 19, no. 2 (2014): 103–132; Michael Hunter, "Robert Boyle and the Early Royal Society: A Reciprocal Exchange in the Making of Baconian Science," *British Journal for the History of Science* 40, no. 1 (2007): 1–23.

52 See, for example: 25 June 1662 (OS), RS JBO/1/66; 15 April 1663 (OS), JBO/1/159; 30 December 1663 (OS); JBO/2/23. Journal Book Original, London. Richard Shortgrave (d. 1676) may have been the Operator, see: Marie Boas Hall, *Promoting Experimental Learning: Experiment and the Royal Society, 1660–1727* (Cambridge: Cambridge University Press, 1991), 27.

53 Entry of 18 June 1662 (OS), RS, JBO/1/66.

54 Entry of 24 June 1663 (OS), RS, JBO/1/194.

55 For a discussion of the early *Philosophical Transactions*, vide: Adrian Johns, "Miscellaneous Methods: Authors, Societies and Journals in Early Modern England," *British Journal for the History of Science*, 33, no. 2 (2000): 165–174.

56 Robert Boyle, "New Pneumatical Experiments About Respiration," *Philosophical Transactions of the Royal Society of London* 5, no. 62 (1670): 2011.

57 Anita Guerrini, *Experimenting with Humans and Animals: From Galen to Animal Rights* (Baltimore: Johns Hopkins University Press, 2003), 38.

58 Boyle, "New Pneumatical Experiments About Respiration," 2024.

59 Ibid., 2025–2026.

postscript that "divers Gudgeons since taken dy'd there in much fewer dayes" suggests that several trials were run. The *Historia piscium* lauds Boyle for his "most excellent experiments" on the effects of water pressure upon bodies of air.[60] It recounts an experiment to fill up a swim bladder with air and submerge it in a clear, deep vessel filled with water. The deeper the bladder was plunged, the more contracted it would become, and vice versa.[61]

Fellows did not only pursue their inquiries on fish within the confines of Gresham College, where their weekly meetings took place.[62] Hooke recounted coming across a porpoise displayed at Ulbars (possibly a fishmonger's shop) in November 1679.[63] He bought the specimen and transported it to Garraway's coffee house, near the Royal Exchange.[64] Here he performed a public dissection.[65] Just like demonstrations of instruments, examinations of animal species in taverns or coffee houses could facilitate discourse on natural phenomena among individuals of various stripes.[66] These might well be people possessing valuable experience regarding the subject, such as sailors. Hellawell has demonstrated, for example, how the Society considered seamen uniquely positioned to record and examine certain natural phenomena. The Fellows asked them to conduct experiments and make observations while at sea, for example recording sightings of species of birds, fish, and other animals as well as magnetic variations of the tides.[67] Her study confirms Lux and Cook's hypothesis that the Royal Society was a relatively open institution that welcomed contributions from outside of its own geographical and social reach,

---

60 *Hist. pisc.*, 8.

61 Similar questions are asked in A.I. and Robert Boyle, "A Conjecture Concerning the Bladders of Air That are Found in Fishes, Communicated by A.I.; And Illustrated by an Experiment Suggested by the Honorable Robert Boyle," *Philosophical Transactions of the Royal Society of London* 10, no. 114 (1675): 310–311. The experiment entailed placing a specimen into a tall, long-necked vessel filled with water, and observe whether upward or downward motions of the fish caused changes in the water level.

62 Michael Hunter, "A 'College' for the Royal Society: The Abortive Plan of 1667–1668," *Notes and Records of the Royal Society of London* 38, no. 2 (1984): 159.

63 Robert Hooke, *The Diary of Robert Hooke, 1672–1680*, eds. Henry W. Robinson and Walter Adams (London: Taylor and Francis, 1935), 430–431. See also Hunter, *Wicked Intelligence*, 118.

64 Rob Iliffe, "Material Doubts: Hooke, Artisan Culture and the Exchange of Information in 1670s London," *British Journal for the History of Science* 28, no. 3 (1995): 286.

65 See: Noah Moxham, "Edward Tyson's *Phocaena*: A Case Study in the Institutional Context of Scientific Publishing," *Notes and Records of the Royal Society of London* 66, no. 3 (2012): 235–252.

66 Adrian Johns, "Coffeehouses and Print Shops," in Park and Daston, *Cambridge History of Science*, vol. 3, 336.

67 Hellawell, "Best and Most Practical Philosophers," 36, 46.

provided, of course, that a member vouched for the credibility of any such informant.[68]

While Hellawell proposes further, specialised case studies be conducted of the evaluation of the knowledge and skills of other occupational groups,[69] she signals that this can be difficult as such groups do not always fit "the conventional artisanal mold."[70] Like seamen, fishermen do not readily fall into those historiographical categories of workmen who have received sustained attention of historians of science over the past decades, notably invisible technicians and artisans. The work of fishermen and fishmongers was, after all, not technical in the sense that they handled (scientific) instruments – in contrast to, for example, those technicians who assisted Boyle.[71] They also do not quite resemble the kind of self-aware artisans that we might encounter in the works of Pamela Smith and Pamela Long. These include the painters, goldsmiths and other artists who produced the many recipes, manuals, drawings, paintings, casts, or ceramics that have come down to us today.[72] This book situates fishermen and fishmongers somewhere in between practical and artisanal communities, allowing for the variety of approaches these individuals adopted in their engagements with fish.

We can get a better view of how the work of fishermen, fishmongers and cooks was regarded through comparing them with workers whose experiences of the natural world stemmed from daily labour both in and on the earth. In her article on (in)visible 'earth workers', Barnett has examined how gentlemen naturalists carved out their own social and epistomological status against that of the quarrymen, shepherds, construction workers and ditch-diggers who supplied these men with fossils and other interesting finds. She argues that by acknowledging the physical labour of these earth workers – even if they seldom did so by name – naturalists showed that they had abstained from

68 David S. Lux and Harold J. Cook, "Closed Circles or Open Networks? Communicating at a Distance During the Scientific Revolution," *History of Science* 36, no. 2 (1998): 201.

69 Hellawell, "Best and Most Practical Philosophers," 33. An example of such a case study are the miners discussed in Kerrewin van Blanken, "Earthquake Observations in the Age Before Lisbon: Eyewitness Observation and Earthquake Philosophy in the Royal Society, 1665–1755," *Notes and Records of the Royal Society of London* (2020), published online ahead of print.

70 Hellawell, "Best and Most Practical Philosophers," 33–34.

71 Steven Shapin, "The Invisible Technician," *American Scientist* 77, no. 6 (1989): 544–563; Rob Iliffe, ed., *Technicians*, special issue of *Notes and Records of the Royal Society of London* 62, no. 1 (2008): 3–148.

72 Pamela O. Long, *Artisan/Practitioners and the Rise of the New Sciences, 1400–1600* (Corvallis: Oregon State University Press, 2011); Smith, *The Body of the Artisan*.

performing the kind of manual work considered unbecoming to their rank.[73] Perhaps the act of fishing, too, tarnished the reputation of the naturalist. This might have well depended, as the heavy work of lifting up nets filled to the brim with fish was something altogether different from the leisurely angler's pastime of reeling in a line. We can also think of handling the slippery bodies of fish, cutting them open like a cook or maid would do.

There is a considerable lacuna of sources when it comes to the attitudes of fishermen and fishmongers towards the study of nature. This scarcity is due to various reasons. The quite obvious one is that fishermen and fishmongers have not generally left much behind in writing – with some exceptions here and there, as we will see shortly. Within practical and artisanal communities, interactions were likely to have been of a local and oral nature. These are precisely the kinds of connections that are difficult to reconstruct, and that tend to be overlooked as a result of the emphasis on texts when reconstructing early modern networks.[74] And yet, as Azzolini has argued, we "accord undue weight to the authority of writers" while not taking the spoken word into account.[75] While we know from the reports of scholars that they had conservations with fishermen and fishmongers, their accounts are of course edited and much condensed. They offer, therefore, only mediated access. As the passage opening this section has also highlighted, the authors and compilers of the *Historia piscium* ultimately selected what was included in the book, and what was left out.

In order to get a less one-sided view of the interactions between Fellows and fishermen, another approach is required: we must consider sources other than the accounts of scholars. Take for example the petition (1663) of the London fishmongers held in the archives of the Royal Society. This petition, which was presented to Parliament, was read aloud during a Society meeting.[76] The fishmongers wished "that our Sea coste & rivers may swarme with the fry & brood of fish, & our Towns and Cittyes better provided for" through stricter enforcement of the law prohibiting too many young fish from being taken.[77] Although the document does not touch upon natural historical reflections explicitly, this remark shows that these fishmongers were occupied with the generation and growth of fish. It also reminds us that, while the relative inconspicuousness of fishermen and fishmongers may lead them to seem like a monolithic group,

73 Barnett, "Showing and Hiding," 261.

74 Ruth Ahnert, "Maps versus Networks," in *News Networks in Early Modern Europe*, eds. Joad Raymond and Noah Moxham, (Leiden: Brill, 2016), 140.

75 Azzolini, "Talking of Animals," 299–301.

76 Petition, RS Cl.P.15i/8, Classified Papers, f2r. It was read on 23 September 1663 (OS) and brought in by John Graunt (1620–1674).

77 Ibid.

they had their own stakes and interests in the world of fish.[78] That these interests need not only be economic becomes clear from the rare manuscripts of the hands of men who caught and traded in fish while also subjecting them to closer study.[79] One such manuscript is that of the aforementioned fisherman and burger Leonhard Baldner, entitled *Vogel-, Fisch- und Thierbuch* [Book of Birds, Fish and Animals]. Willughby and Ray cited it throughout the *Historia piscium*. As it offers a unique entry into Baldner's own ideas about what the study of fish entailed, the last section of chapter will discuss this work in more detail.

So, while this chapter departs from the *Historia piscium*, and asks how the compilers of the work incorporated the experiences of fishermen and fishmongers, it also considers the perspective of the latter's groups where possible. This can give us a more well-rounded idea of what such exchanges may have entailed. We will, for the remainder of this chapter, reconstruct the nature, extent, diversity and significance of the contributions of practical men to the *Historia piscium*, and how these were evaluated by the Fellows.

2 Knowledge at the Fish Market

Fishermen take centre stage in the engraved title page, even if they are depicted as rather more genteel individuals than they probably were. Fishermen and fishmongers provided (if not always wittingly) the raw material for natural historical and philosophical investigations. Fellows considered access to fresh specimens of fish to be of great importance. This section compares the kinds of evidence that could be taken from preserved specimens, illustrations and fresh specimens. It thus picks up on themes such as field trips, illustrations, and natural historical collections that have been mentioned in Chapter 1 but thus far have not been elaborated on.

At his home in Middleton Hall, Willughby could examine the plants in his garden and the animals in his *vivarium*,[80] just as Ray examined the trees in his

78 An overview of the various types of fishing in different European regions can be found in A.R. Michell, "The European Fisheries in Early Modern History," in *The Cambridge Economic History of Europe*, eds. E.E. Rich and C.H. Wilson, (Cambridge: Cambridge University Press, 1977), 133–184.

79 A striking example for the sixteenth century is the manuscript entitled *Visboeck* (Fish book) by Adriaen Coenen (1514–1587). For more on him, see Egmond, "On Northern Shores," 132–139.

80 Poole, "The Willughby Library," 229.

own orchard.[81] When Willughby and Ray travelled through the British Isles and across continental Europe, they frequented markets to get their hands on new species of birds and fish. As Ray put it, they "visited almost all the chief fishing ports of England, and the markets of Belgium, Germany, Italy and France; [...] bought all the species new to us and described them so that the reader can easily recognise them."[82] Their daily visits to the fish market in Rome produced rich results, as they found that there was "scarce any fish to be found anywhere on the coast of Italy but some time or other it may be met withal heer [*sic*]."[83] Travel companion Philip Skippon listed no fewer than eighty-nine species of fish that they had come across at Venice's market.[84] He described, for example, finding "a little fish with a scarlet belly, called Sanguinuole" in the market of Brescia.[85] Visiting (fish) markets to spot new specimens was in fact a widely utilised practice. When stationed in Jamaica in the service of the second Duke of Albemarle, for example, the physician and collector Hans Sloane scoured the island's markets for new specimens to examine.[86]

The piscine wealth to be found at fish markets was further proof that the underwater world teemed with creatures that merited closer examination. In one of his physico-theological treatises, Ray marvelled – echoing psalm 104.25 – "The Sea, what infinite Variety of Fish doth it nourish!"[87] While fish were indeed wonderfully varied, Ray also believed that God had created a fixed number of species.[88] From the onset, Ray set his expectations for the *Historia piscium* at a high mark. As he wrote to Robinson in 1684: "For this history of fish, I can warrant it to be as full and perfect as to the number of species, and their descriptions [...] as was the history of birds."[89] As the previous chapter discussed at length, Willughby and Ray's idea of a perfect fish book differed from those extensive volumes full of anecdotes, fables and proverbs, which certain earlier Renaissance authors had compiled. Rather, they wished to rectify the unnecessary duplication of species by plotting characteristic marks.

81 Dániel Márgocsy, *Commercial Visions: Science, Trade, and Visual Culture in the Dutch Golden Age* (Chicago: Chicago University Press, 2014), 53.
82 As quoted in Raven, *John Ray*, 365.
83 Ray, *Observations*, 362.
84 Skippon, *Journey*, 496, and Kusukawa, "*Historia Piscium* (1686) and Its Sources," 323.
85 Skippon, *Journey*, 571.
86 James Delbourgo, *Collecting the World: The Life and Curiosity of Hans Sloane* (London: Allen Lane, 2017), 91.
87 Ray, *The Wisdom of God*, 78.
88 Kusukawa, "*Historia Piscium* (1686) and Its Sources," 313.
89 Ray, letter to Robinson, 13 March 1684 (OS), *Correspondence of John Ray*, 164.

This was also a matter of precise language, as the previous chapter explained. Their study of fish, and of nature more generally, was carried out in the context of larger philosophical reflections on the connections between knowledge and language, an interest they shared with fellow Royal Society member Wilkins. Along with many of his contemporaries, Wilkins thought that God had greatly compromised man's ability to communicate in his judgement that followed the attempt to build the tower at Babel (Genesis 11.1–9).[90] Wilkins therefore set out to compose a universal language, by creating word tables that showed the true relation between words and things. Willughby and Ray both contributed to Wilkins' project, which eventually appeared as *An Essay Towards a Real Character, and a Philosophical Language* (London, 1668).[91] Ray, however, would later privately admit to be "ashamed and disgusted" to have been so publicly associated with a project that he, found, at its core, to be ludicrous.[92] This was not because he disagreed with the idea that a proper connection could – and should – be established between a word and a thing: he himself was very much concerned, as we will see, with reconciling the proper relations between fish species and their names. Ray shared Wilkins' quest for a language that was stripped of ambiguity, especially when it came to describing living things.[93] What he denounced, however, was the imposition of a pre-contrived system onto nature's rich variations. By way of contrast, Ray was convinced that true knowledge came from the senses.[94]

When deploying the senses to study a species of fish, having recourse to (more or less) fresh samples was much to be desired. For this, they need not always visit fish markets, as sometimes fishermen delivered specimens to the naturalist's doorsteps. In a letter to the Royal Society detailing his dissection of a porpoise, Ray related how, during his visit to Wilkins in Chester in late April 1669, he had had "the good fortune to meet with a young porpess of a convenient size for dissection, brought tither by some fishermen, who caught

90 Kusukawa, "The *Historia Piscium* (1686)," 183. A standard work on the natural philosophical pursuits of constructing a universal language is Mary Slaughter's *Universal Language and Scientific Taxonomy in the Seventeenth Century* (Cambridge: Cambridge University Press, 1982).

91 Ray worked on the plants, Willughby on the animals. John Wilkins, *An Essay Towards a Real Character, and a Philosophical Language* (London: S. Gellibrand and John Martyn, 1668).

92 David Cram, "Francis Willughby and John Ray on Words and Things," in Birkhead, *Virtuoso by Nature*, 255.

93 See also: Mary Slaughter's *Universal Language and Scientific Taxonomy in the Seventeenth Century* (Cambridge: Cambridge University Press, 1982), 62–64.

94 Kusukawa, "The *Historia Piscium* (1686)," 184; the role of sensory experience in the Royal Society is discussed at length in Wragge-Morley's *Aesthetic Science*.

him upon the sands, where the tide had left him [...]."[95] These men seemed well aware that the novelty value of certain fish washed ashore could be converted into actual coin. Their hustling was rewarded; the bishop purchased the animal (for an unknown sum) and handed it to Ray for dissection.[96] As was shown in the previous chapter, examining animals' anatomies was in fact a key component of Willughby and Ray's research; the dissection of a flair was an exemplary piece of the kind of close observation that they held up as an ideal.

When no fresh specimen was at hand, they made do with preserved ones. Willughby and Ray were dependent on what their correspondents were willing and able to send them, or what they could find or buy themselves. Willughby himself amassed a collection of "Birds, Fishes, Shells, stones and other fossils, seeds, dried plants, coins, etc" on his estate.[97] In London, dried fish could, as we have learned in the introduction to this chapter, even be found in taverns. And as discussed in Chapter 1, the Royal Society itself possessed a repository of objects. The catalogue made from it included a section on aquatic fauna entitled "Of Fishes" encompassing the "RIB of a TRITON, or MAREMAN" alongside several kinds of whale bones, the horn of a sea-unicorn that the Icelanders called a narwhal, some seals, the claw of a lobster – all of which attests to the wide category of creatures the word 'fish' continued to encompass in this period.[98] The collection may have included a great range of species, but its value for making proper species descriptions was limited, because, as Michael Hunter has noted: "preserved exhibits were decidedly inferior to live ones".[99]

The difference in utility between that of a living specimen and a dead, prepared one was especially marked in fish because they disintegrated so easily. What is more, different species demanded different methods of preservation. Larger specimens would often be dried, and sometimes stuffed with hay so as to retain some of their shape. Smaller specimens were usually stored within glass jars filled with spirits. Each method of preservation had its merits and

95 John Ray, "An account of the dissection of a Porpess, promised numb. 74; made, and communicated in a letter of Sept. 12 1671, by the learned Mr. John Ray, having there in obser'd some things omitted by Rondeletius," *Philosophical Transactions of the Royal Society of London* 6, no. 76 (1671): 2274.

96 *Hist. pisc.*, 32.

97 Poole, "The Willughby Library," 230; a part of this collection is still extant, see: Tim R. Birkhead, Paul J. Smith, Meghan Doherty and Isabelle Charmantier, "Willughby's Ornithology," in Birkhead, *Virtuoso by Nature*, 277; Charmantier, Johnston and Smith, "The Legacies of Francis Willughby," in idem, 375.

98 Grew, *Musaeum Regalis Societatis*, 81–124.

99 Michael Hunter, *Establishing the New Science: The Experience of the Early Royal Society* (Suffolk: The Boydell Press, 1989), 152.

pitfalls; inundating specimens with spirits, for example, was rather costly and not altogether attractive for display, whereas dried specimens could become brittle so that only the sturdier parts of the fish endured.[100] Objects preserved in the latter fashion also failed to allow for any examination of internal organs. These parts of the fish would be removed along with the flesh during the process of preservation as, unlike the fish's skin, the internal organs would not desiccate easily. Regardless of the preservation strategy used, however, the fish in question would often lose much, if not all, of its original colour in the process.

That there was often a considerable discrepancy between a fresh specimen of a species on the one hand and a preserved exemplar on the other was far from lost on the Fellows of the Royal Society. Grew had written in his description of a "little SEA-UNICORNE [...] sent from Brasil", not earlier described or depicted, that from the top of the fish "is prolonged a smooth (now) blackish, round, taper'd, strait Horn [...]" and that the fish itself is "cover'd with a (now) blackish, thick and tough Skin, and when you draw your hand forward, also rough."[101] The insertion of "(now)" shows that Grew was cognizant of the fact that the passage of time probably had affected the look of the specimen since it had made its way over to England from the South Americas. This caveat was included into the species description of this '*Monoceros Minor Mus. Soc. Reg.* D. Grew' in the *Historia piscium*. In the description of the horn, it is noted that the blackish colour could be glimpsed "in exsiccato pisce", *viz.* in the dried fish in the Repository of the Society.[102]

Images could address this problem of deterioration – at least, to an extent, as we have also seen in the previous chapter. The importance of illustrations for the *Historia piscium* was signalled on its engraved title page by the inclusion of the helmeted artist. As Chapter 1 has discussed at length, the book included new figures that were usually based on drawings that Willughby, Ray or others in their circle had acquired. While some preliminary sketches of fish made during their trip seem exist among the Middleton collection, these did not make it into the book.[103]

100 See also: Peter Davis, "Collecting and Preserving Fish: A Historical Perspective," in MacGregor, *Naturalists in the Field*, 149–165; Marlise Rijks, "Scales, Skins, and Carapaces in Antwerp Collections," in Marjolein Bol and Emma C. Spary (eds.), *The Matter of Mimesis: Studies on Mimesis and Materials in Nature, Art and Science* (Leiden: Brill, 2023), 295–320.

101 Grew, *Musaeum Regalis Societatis*, 104; the interjection 'now' is found in other object descriptions too, among others on 98, 100.

102 *Hist. pisc.*, 216.

103 Kusukawa, "*Historia Piscium* and Its Sources," 321.

One of the sources for illustrations was a manuscript, now inscribed 'A Book of Fishes done at Hamburgh, with Mr Ray's Notes', which has hitherto received scant attention from historians.[104] As of yet, very little is known about how it came into Ray's possession, or even when or where it was produced. During their tour through continental Europe in the mid 1660s, Ray and his company had not ventured further north in the German states than Cologne, so he must have acquired it elsewhere than in Hamburg. The manuscript contains dozens of coloured illustrations of aquatic fauna, executed in watercolour and gouache. These illustrations are accompanied by cursory descriptions in a German hand, which appears to be from the sixteenth century.[105] Certain drawings in the manuscript show an unmistakeable correspondence to a set of fish drawings within the Gessner-Platter albums recently discovered by Florike Egmond;[106] these are clues that can help throw light on the manuscript's origins. Ray's annotations give insight into how he used the book. He commented, for example, on the correct identification of a species ("these are not separate species, but the front and back side of the same fish") or on the quality of certain drawings ("badly painted").[107] While, as Chapter 1 also suggested, the natural historical value of illustrations was related to the skill of the artist and the freshness of the specimen concerned, and while the former might have been reasonably simple to ascertain, the latter would remain difficult had one not personally seen a suitably lively, or at least fresh, example of the species. Fish tend to change appearance soon after being taken out of the water, and Leah Aranowsky has argued for drawings of dead fish that they reflect the interstitial time between life and death, observation and presentation.[108] The qualifying phrase "drawn from the life", with its multivalent early modern usages, thus takes on special meaning in the case of fish.

Both preserved objects and drawings, therefore, came with their own limitations for representation. This was potentially problematic, as we saw in the previous chapter, as meticulous attention to detail was highly desirable if fish were to be properly distinguished from one another. A characteristic

---

104 Anonymous, *A Book of Fishes done at Hamburgh, with Mr Ray's Notes*, Add MS 5308c, British Library (hereafter BL), London; the manuscript stems from Sloane's collection, and I thank Sachiko Kusukawa for drawing my attention to it.

105 The watermarks in the paper, furthermore, date to the mid-sixteenth century.

106 Gessner-Platter Albums, UBA, hs. III C 22. On this album, see Egmond, "A Collection within a Collection," 149–170.

107 'Non sunt distincta species, sed ejusdem piscis pars supina & prona' BL, Add MS 5308c, f2v; 'male pingitur' ibid., f5v.

108 Leah Aranowsky, "On Drawing Dead Fish," *Environmental History* 21, no. 3 (2016): 549.

mark might well be lost in the preservation process, or inadvertently left out of a drawing. It is probably for these reasons that the experiences of fishermen and fishmongers were particularly handy. They saw, after all, a relatively large quantity of each species of fish, and live examples at that, as opposed to either the few dried exemplars in natural historical collections or possibly imprecise drawings that were available to naturalists. As the following section will show, the larger 'sample size' of specimens that these fishermen had observed proved useful for Willughby and Ray for drawing conclusions about demarcating species.

Before fish could be captured on paper, they first needed to be caught. One can easily forget this when looking at the engraved plates in the *Historia piscium*, which present the fish as if untouched by human hands, exhibiting none of the tell-tale marks left by hooks or nets.[109] One exception to this rule is the engraving of a species of flatfish which does convey obvious traces of its capture: a thin black cord has been tied from its head to the peduncle of its tail.[110] The engraving was based on one of the drawings (Figure 7) in the 'A Book of Fishes done at Hamburgh.'[111] This particular manner of tying up flatfish is depicted in various fish still lifes by seventeenth-century Netherlandish painters such as Abraham van Beijeren (1620–1690), Isaac van Duijnen (1628–*c.*1680) and Jacob Foppens van Es (1596–1666). These still lifes often show fish specimens acted upon in one way or the other: they are cut, sliced, smoked or tied. This way of binding a flatfish head to tail seems to have served a very practical purpose, namely to facilitate its transport, or delay the spoiling process.[112] The illustration serves as a reminder that fish had to be caught, carried, stored and preserved before they could be subjected to scrutiny; and thus were subject to the attentions of many individuals, fishermen, fishmongers and other handlers, before they could be subjected to the gaze of the naturalist.

109 Cf. the drawing of a spiky blowfish that Gessner had made, including a hook and tasseled string, and its printed counterpart in *Hist. anim.* IIII, 155, where these have not been represented, although a trace is still visible through a slight bump on the body. See: Egmond, *Eye for Detail*, 160–163.

110 *Hist. pisc.*, tab F1.

111 BL, Add MS 5308c, f4v.

112 Julie Berger Hochstrasser, "From the Waters: Fish Still Life," in *The Magic of Things: Still-life Painting, 1500–1800*, ed. Jochen Sander (Berlin: Hatje Cantz, 2008), 188; "Description of Isaac van Duynen's 'Stilleven met vissen op een tafel'," *Hoogsteder Journal* 3 (1997), 21.

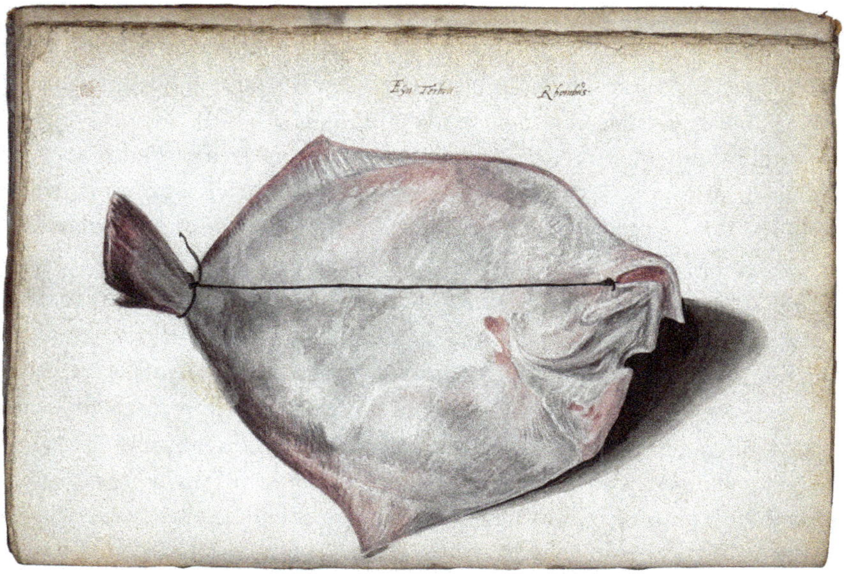

FIGURE 7 Drawing of a species of flat fish. MS 5308c f4v, The British Library

3 Detail and Distinction

Fishermen did not only supply the goods for natural historical research, but were also sources of knowledge in themselves. For Willughby and Ray, the fishermen embodied several different types of evidence, all of which could be recorded. First of all, fishermen shared the techniques they used to catch the fish. Willughby's and Ray's interest in these techniques is evident from some of the species descriptions in the *Historia piscium*, which explain the intricacies of catching herring or trapping tuna.[113] The latter is even rendered in one of the few diagrams in the book, which elucidates the ingenious system they saw in Marseille. When in Sicily, Ray and Skippon took the opportunity to examine fishing from up close. In his travel account, Ray related that they had hired a boat so that they could better understand how swordfish were caught.[114] While they did not witness the capture of any such fish, they did take this opportunity to study the harpoons that the fishermen had brought along for the occasion. Similarly, the *Ornithology* was furnished with several pages expatiating the art

113 *Hist. pisc.*, 220 and 178.

114 Greengrass, Hildyard, Preston, Smith, "Science on the Move," 183.

of fowling.[115] This attention to techniques for catching and trapping animals is on par with the broader interest of the Fellows in various trades, for which the Royal Society set up an official program.[116] Their occasional notes regarding the taste of certain fish can also be read in this light.

Secondly, Willughby and Ray recorded common words in various languages and dialects during their travels through the British Isles. For example, when visiting the West Country of England in 1667, they noticed that Cornish differed only a little from Welsh and also that it was much akin to the Breton language. The similarities were such "that they [the Cornish and the Bretons] understand one another, as we found by severall Fisherman of that countrey w[hi]ch were then drying of cartilagineous Fish at Pensans & St Ives."[117] These fishermen, then, shared knowledge about which words were used for what things in different regions. This was not tangential to Willughby and Ray's project. In fact, being attentive to the words for fish in various dialects was key to their ambition to bring order to the world of fish, as will be elucidated below.

Last not but not least, fishermen offered invaluable insight into the occurrence of species. When Ray toured through the British Isles in 1662 with Willughby, he compiled catalogues of English birds, fish, metals and minerals.[118] He noted down several fish taken around Pensance and Saint Ives in Cornwall, presented to him by "one of the ancientest and most experienced fishermen", who remains nameless.[119] Ray here stressed his informant's decades-worth of experience; other Fellows used similar phrasing while asserting the seniority of the seamen they had consulted.[120] The first entry on Ray's fish list was a whale, which the old fisherman had spotted from the coast. Ray added that he could not tell them of what sort it was, remarking that "*vulgus enim non distinguit*" – the common people, after all, do not distinguish.[121] In the *Historia piscium* it was similarly declared that fishermen do not really discern the mackerel from

115 This had not been part of the Latin original; Birkhead, Smith, Doherty, and Charmantier, "Willughby's Ornithology," 283.

116 Kusukawa, "*Historia Piscium* (1686) and Its Sources," 329. For this interest, see: Kathleen H. Ochs, "The Royal Society of London's History of Trades Programme: An Early Episode in Applied Science," *Notes and Records of the Royal Society of London* 39, no. 2 (1985): 129–158.

117 John Ray, *Further Correspondence*, 262–263.

118 John Ray, *A Collection of English words, not generally used ... in two alphabetical catalogues, ... northern ... [and] southern counties, with catalogues of English birds and fish, and an account of preparing ... metals and minerals* (London: Thomas Burrell, 1674).

119 Ibid., 97.

120 Hellawell, "The Best and Most Practical Philosophers," 44.

121 Ray, *A Collection of English Words*, 97.

any other fish that may look like it.[122] These men's seeming lack of interest in the categorisation or classification of fish ran very much in opposition to Willughby and Ray's asserted aim, namely to precisely distinguish between species.

Ray's remark was somewhat unjust. Not only did the diversity to be found in fish present a complex puzzle, as species often closely resembled each other and could thus only be differentiated through subtle variation, but fishermen's distinctions were also important for Ray as he attempted to try to solve such conundrums. Consider the following passage, in which Willughby and Ray deliberate on whether sprats formed a separate species or were nothing else than the offspring of herring:

> A certain senior fisherman from *Cornwall*, whom we have consulted about this matter and other things, has told us that two kinds of *Sprats* are caught in the sea which flows near to Cornwall, one of Herring, another of Pilchards or the offspring of Celerini, which can in turn easily be distinguished from another. Pilchards frequent the shores of Cornwall and Devon, they very rarely progress further to the east in the British sea; from whence elsewhere around England only one type of Sprat is found.[123]

Here, yet again, a fisherman – possibly that same wise and experienced individual – imparted his knowledge. His answers did not make matters simpler, as he explained that there are, in fact, different kinds of sprats, which stem from at least two different species, and that these are, furthermore, not distributed equally along the coastlines of the British Isles. A looming problem in these interactions was that a fish might be accorded one name in Cornwall, and yet another in London. The 'Scad' in Cornwall was known as a 'horse Mackrell' in London; conversely, the species of flatfish that Londoners dubbed a 'Pearle', the Cornish called 'Lug-aleaf.'[124] In keeping with Willughby and Ray's preoccupations with language, the *Historia piscium* and its related writings abound

122 *Hist. pisc.*, 182.

123 'Piscator quidam senior *Cornubiensis*, quem super hac re aliisque consuluimus, nobis retulit duo *Sprattorum* genera in mari Cornubiam alluente capi, alterum Harengorum, alterum Pilcardorum, seu Celerinorum sobolem, quae à se invicem facile distingui possint. Pilcardi Cornubiae & Devoniae littora frequentant, ulterius in mari Britannico orientem versus raro progrediuntur; unde alibi circa Angliam unicum tantum Sprattorum genus invenitur.' *Hist. pisc.*, 221.

124 The names 'Scad' and 'Lug-Aleaf' are those listed in the species descriptions in *Hist. pisc.*, on page 290 and 95 respectively; the 'horse Mackrell' and 'Pearle' are handwritten

with attempts to establish which fish was called by which name where, and by whom.

The taxonomies of fishermen did not always overlap with those of the naturalist. This added a linguistic layer to the already intricate puzzle presented by the relationships between the various species. Ray wrote to Lister: "Of the flat cartilaginous [fish] I have seen and described four or five sorts, but I am to seek what our fishermen mean by the Skate, and what by Flair, and what by Maid – as Skate-maid, Homelyn-maid, Thornback-maid, &c. &c."[125] Distinctions between (or even within) species by people that engaged with fish in a more practical sense also appear to have been based on attributes with particular relevance to their commerce. In the species description of the herring, it is explained that the people who washed, salted and dried this fish, and who were called Towers, separated it "into six species or rather grades".[126] These encompassed the 'fat herring', which was large and fat, and the 'meat herring' which was equally large and rich in meat but less fat.[127] 'Pluck' was the name used for herring damaged or torn from being stuck in the nets, while a 'shotten herring' had emptied itself of its roe.[128] We thus find, subsumed in Willughby and Ray's natural historical taxonomy based on characteristic marks, a further taxonomy drawn up from properties stemming from commercial practice.

Ray's erstwhile fellow Cambridge student and vicar of Brignall Ralph Johnson (1629–1695) wrote to complain of how difficult it was to decide whether dissimilar-looking exemplars of salmon were truly different species, or rather one and the same species at different stages of growth.[129] He said that in "the mouth of Eden in Cumberland the fishers have four distinctions of yearly growth (after the first summer, when they call them free, or frie, as we smowts, or smelts) before they come to be lackes; and this, they say, they have curiously observed, by fixing so many pins in the fins of yearlings, or two years old, and after taking them again; [...]."[130] This procedure, of fixing pins into individual

additions in the Royal Society's copy of Willughby and Ray, *Historia piscium*, RS, RCN 18574, on the pags mentioned (both TR).

125 Ray to Martin Lister, 19 December 1674 (OS), *Correspondence of John Ray*, 113.

126 '[i]n sex autem species seu potius ordines [...]' *Hist. pisc.*, 220. The translation comes from Hans Aili and Theodore W. Pietsch, *Peter Artedi: Reformer of 18th Century Zoology* (Stockholm: Stockholm University Press, 2024), 82.

127 Ibid., 82–83.

128 Ibid., 83.

129 Johnson also shared observations on and specimens of birds and plants. Raven, *John Ray*, 319. A biographical note can be found in *Teesdale Record Society* 15 (1945): 9–32.

130 Ralph Johnson to Ray, 16 April 1677 (OS), *Correspondence of John Ray*, 127.

specimens and tracing their growth over a period of time, was effectively an experiment. Like the experiments conducted by the Fellows, it was designed to allow for certain observations to be made. Fishermen's distinctions between salmon of different ages were deemed dependable enough to be included into the book:

> And what is handed down by authors about the quick *growth* of small salmon in the sea does not find faith with us: for our fishermen distinguish salmon by each year of their age, as we have said above, and they say that they are not full-grown before the sixth year of their life.[131]

Willughby and Ray here plainly stated that they placed their trust in the collective account of 'their' fishermen rather than in the written knowledge transmitted by various earlier authors (whom they do not specify here). This sentence can also be read as a rhetorical phrasing reminding the reader that relying on ancient authors is a matter of faith, whereas believing the fishermen is a matter of lived experience.[132]

How could one tell whether a specimen was exemplary for its species? Fishermen and fishmongers had a good sense of irregularities and averages. A fishmonger, for example, told Willughby and Ray that bigger specimens of salmon weighed around six pounds.[133] Furthermore, the Cambridge fishmonger Mr. Mayfield, who went down to the London market every Friday to procure species not readily available in his own town, shared worthwhile observations.[134] The physician Peter Dent (*c.*1628–1689) wrote to Ray that "Mr. Mayfeild [*sic*] could not procure any dried *Mayds* or *Thornback* at the mart. He helped me to a fresh *Thornback*, which he said was full grown: its weight was ten pounds."[135] Dent added the fishmonger was "acquainted with the Tamworth carrier and will undertake to send you any of these [fishes] fresh into the country [...]",[136] and thus could also do deliveries. He had furthermore told Dent that he once sold an exceptionally large specimen of flair to the cook

131 'Quae de celeri Salmunculorum in mari *auctu* ab Autoribus traduntur apud nos fidem non inveniunt: nostratis enim piscatores Salmones annuatim ab aetate distinguunt, ut superius diximus, neque ante sextum aetatis annum perfici aiunt.' *Hist. pisc.*, 192.

132 I thank Pete Langman for this observation.

133 *Hist. pisc.*, 196.

134 Raven, *John Ray*, 393.

135 Peter Dent to Ray, 15 February 1674 (OS), *Correspondence of John Ray*, 15–17.

136 Dent to Ray, 15 February 1674 (OS), passage omitted in Lankester's *Correspondence of John Ray* but reproduced in Gunther, *Further Correspondence of Ray*, 113.

of St John's College in Cambridge, and it ended up feeding all those attending lunch that day. Dent sought verification of the story from the cook in question, and having received it, he passed it along to Ray who then inserted it into the *Historia piscium*.[137] The reader could rest assured that the fishmonger Mayfield was of trustworthy character [*fide dignus*].[138]

Fishermen and fishmongers could furthermore tell whether a certain specimen was male or female, and how particular species procreated. The dependable Mayfield, for example, assured Dent that flairs were viviparous.[139] While Dent doubted whether this was true, he resolved to observe the alterations of the fish's eggs on a weekly basis and give Ray a full account.[140] Although Dent's ultimate findings cannot be found in Ray's correspondence, the letter underscores the fact that the statements of fishmongers, like that of fishermen, merited further research and that their claims invited both validation and repudiation.

The *Historia piscium* frequently cited from Leonhard Baldner's manuscript *Vogel-, Fisch- und Thierbuch*, mentioned earlier in this chapter as a suitable source to reconstruct the experiences of those who worked with fish on a daily basis. Baldner was the first fisherman mentioned by name in the *Historia piscium*; rarer still, his portrait has come down to us.[141] This is probably because Baldner was not a 'typical' fisherman. He was born into an established Strasbourg fishing family (whose crest consisted of three crossed fish), must have received some education as he could read and write, and combined his occupation as fisherman with a seat in the city council.[142] Baldner produced several, largely similar, manuscripts in quarto describing the quadrupeds, birds, fish and insects of his home region, most of which were skilfully illustrated and painted by the painter Johann Georg Walther (1634–1697). While some of these manuscripts have sadly been destroyed or lost, 4 copies are

137 Where it was now claimed that the flair had fed all College's hundred-twenty alumni. Kusukawa, "*Historia Piscium* (1686) and Its Sources," 331 and *Hist. pisc.*, 69.

138 *Hist. pisc.*, 69.

139 Dent to Ray, without date, *Correspondence of John Ray*, 120.

140 Ibid.

141 The portraits are at NUL, Mi LM25/80 and Brown University Library, RARE 3-SIZE QL41 .B3 1653 v.1; see also: Kusukawa, "*Historia Piscium* (1686) and Its Sources," 320.

142 Hans-R. Fluck and Albert Scharbach, "Leonhard Baldner – Zu seinem Testament and Nachlassverzeichniss," *Revue d'Alsace* 142 (2016): 293–294. He also collected duties on the Rhine: Armin Geus, "Leonhard Baldner: A Strasbourg Fisherman," *Isis* 55, no. 2 (1964): 196.

known to be preserved in libraries and archives.[143] Both the descriptions and the drawings in these manuscripts as of yet await detailed analysis, and a comparison between the extant editions would be most welcome to offer insight into Baldner's approaches to the study of nature as well as how, through these diligently produced works, he presented himself as a naturalist.

This chapter focusses on the copy in the British Library. Willughby seems to have bought this manuscript, the preface to which is dated 31 December 1653, from Baldner himself during the continental tour.[144] It contains very fine watercolours, and the descriptions are carefully calligraphed; certain details of both the text and the images have been accentuated with gold. Willughby and Ray used the manuscript as a source for their studies of both birds and fish: the *Ornithology* contains 37 drawings from Baldner (making up a little over a tenth of the total illustrations in their work),[145] whereas the *Historia piscium* includes 25 of Baldner's illustrations and cites from it in several species descriptions.[146] We will now discuss how these English naturalists used the manuscript, and what Baldner's own intentions for it were.

In the preface to the *Ornithology*, Ray expressed his appreciation of the high quality of the manuscript's illustrations, praising their great exactness and excellent hand.[147] It struck him that Baldner had taken and described these fish himself, and had them drawn at his own charge and cost. Such curiosity, Ray thought, was "much to be admired and commended in a Person of his Condition and Education."[148] He also acknowledged that he had received "much light and information from the Work of this poor man", which had enabled him to "clear many difficulties, and rectifie some mistakes in *Gesner*."[149] Ray furthermore wrote to Robinson: "though it is not to be supposed, that a man of his education should be able to describe animals well, yet so much might be gathered from the notes he gives, as might lead an understanding

143 Besides the aforementioned copies at the British Library and Brown University, Library, the University Library of Kassel and the Bibliothèque Nationale et Universitaire de Strasbourg both hold a copy. See also: Birkhead, *The Wonderful Mr. Willughby*, 101–103.

144 Leonhard Baldner, *Vogel-, Fisch- und Thierbuch* [Book of Birds, Fish and Animals], BL, Add MS 6485.

145 Birkhead, Smith, Doherty, and Charmantier, "Willughby's Ornithology," 295.

146 Kusukawa, "*Historia Piscium* (1686) and Its Sources," 320.

147 Willughby, *Ornithology*, preface, sig. A6v.

148 Ibid.

149 Ibid. Ray did not read German, and used Frederick Slare's abridged translations of the species descriptions, BL, Add MS 6486, ff12r–23v. That Ray also engaged directly with Baldner's manuscript is evidenced by the Latin names he added to some of its descriptions.

FIGURE 8 Watercolour of a species of carp, inscribed 'Cyprinus' in Willughby's hand. NUL Mi LM 25/51, University of Nottingham Library Manuscripts and Special Collections

and attentive man into the knowledge of them, and with the figures (which are in all very exact) give him so much light as to enable him to determine the species."[150]

On the title page of his manuscript, Baldner proclaimed that both the species descriptions and illustrations conformed to nature.[151] Looking at a drawing that Willughby purchased from Baldner alongside the manuscript, a watercolour of a carp (Figure 8), one can see why Ray was so enthused.[152] The artist has drawn the fish from a slight bottom perspective view, and diligently rendered the scales, and fins, which in particular show fine brushstrokes. By subtly applying a greyish light blue pigment to the edges of the gills and scales, a technique known as heightening, the artists conveyed the glistening of a fish that has just been taken out of the water. The drawing was used for the *Historia piscium*.[153] Baldner's intention was that the descriptions and images in his manuscript would complement one another. He stated, for example, that "[t]he species of 'Rothaug' are not dissimilar to that of the 'Rotel', but they are more beautiful in colour and of more rubescent eyes, and fins, as can be seen

150 Birch, *History of the Royal Society*, vol. 4, 390.

151 'Recht Naturliche Beschreibung Und abmahlung [...],' BL, Add MS 6485, f1r; inserting the word "recht", Baldner modestly says they are "almost" natural.

152 NUL, Mi LM 25/51.

153 *Hist. pisc.*, Tab Q1. The other loose drawings are a perch (Mi LM 25/58) and a portrait (Mi LM 25/80). The former is represented in Kusukawa, "*Historia Piscium* (1686) and Its Sources," 321.

from the illustration [...]."[154] In their description of the 'Rootaug', Willughby and Ray used the same distinctive marks.[155]

The authors looked to Baldner's manuscript for a wider range of observations, copying, for example, his statements on whether a certain species was rare or common, how its appearance could vary along with time or place, when and how it procreated, and when it was best to eat, in the descriptions of no fewer than twenty species.[156] To focus on only those parts of the manuscript that were included in the *Historia piscium*, however, is to miss out on Baldner's own questions and approaches in studying fish. Among the volume's fascinating observations is his account of having caught a sturgeon of "about the thickness of a man", and subsequently finding its bowels to weigh 130 pounds.[157] Like Willughby, Ray and their peers, Baldner thus dissected fish and studied their internal anatomies; he even counted the thousands of eggs in the roe of pike and burbot.[158] He noticed that the species of wood trout took on the colours of their environment: they turned completely white when placed in a white tub, and black once put in a black tub.[159] He disagreed with Gessner that carp were (sometimes) born from mud, and said that they all came from roe.[160] On the whole, Baldner's manuscript shows that he aimed to discern species from one another, to examine their anatomies, to understand how they behaved and procreated, and that he compared the reports of earlier authors with his own observations – again, much like Willughby and Ray.

The preface to Baldner's manuscript gives us a sense of how he envisioned his work. It reveals that the author thought there to be no better place to contemplate God's omnipotence than on and near the water. Since God had at the beginning created the great whales, fish had received His first blessing; and He had also called upon the fishermen to follow him. God had, furthermore,

154 'Die Rothaugen sehen den Rottlen nicht ohngleich, seind aber von farben hüpscher, und Rothere Augen, und Schwümfedern, wie von dem abgemahlten zu sehen [...].' Add MS 6485, f135v.

155 *Hist. pisc.*, 249. Some confusion around the identification of this species is related in Birch, *History of the Royal Society*, vol. 4, 390.

156 Baldner's manuscript is referenced on the following pages: *Hist. pisc.*, 105–107, 118, 201, 125, 227–228, 236, 238, 248, 249, 250, 252–254, 259, 260–262, 265, 266.

157 Add MS 6485, f119r.

158 *Hist. pisc.*, 201 and 125, cf. Add MS 6485, f121r and f134r.

159 Add MS 6485, f125r.

160 Add MS 6485, ff121v–122r. In his German history of fishes, a loose and much abbreviated translation from the Latin, Gessner copied Rondelet's statement that carp are sometimes born from chaos and dirt, and sometimes from seed and roe, see: Gessner, *Fischbuch*, 164–165.

made the rivers of the Rhineland with their endless benefits to those who lived around them. It was this delight in and admiration for the Creation, Baldner submitted, that had inspired him to make this manuscript brimming with the animals that swam, flew and crept in these waters. All of the creatures described in it, he wrote, he had held in his own hands. Each of the species was drawn from life, called by its name, and after sustained study, described briefly from Baldner's own 'experience' [*Erfahrung*].[161] He admitted his attempts were necessarily 'simple' [*einfältig*] and 'scant' [*gering*], casting himself as a modest fisherman and hunter, and bade those considering themselves better suited to write such a work to keep that humble background in mind.[162] At the same time, he emphasised his three decades worth of experience with fish – although he used the terms 'learned' [*erlernt*] and 'studied' [*studiert*] to describe this involvement.[163] Quite apart from its complicating of certain assumptions about what constitutes 'a' fisherman, Baldner's manuscript also testifies to the fluid boundaries of theoretical and practical engagements with nature.

The examples listed in this section offer an idea of the topics Willughby, Ray and their friends discussed with practical men: from distinguishing between species to noting their various names in different languages, the intricacies of procreation to deciding if a certain specimen was of a typical size for its species. The preoccupation of fishermen and fishmongers with, for example, the occurrence of certain species or the growth stages of young salmon, can likely be traced back to commercial considerations, but that was not necessarily the sole motivation. Baldner's manuscript presented a natural historical study in its own right. His book contained observations that are of a practical nature, like whether a certain species is edible, but also included reflections on long-standing theoretical debates into the generation of fishes. He presented himself as a student of nature who did not strive for profit, but instead wished to praise God through studying His Creation. This approach compared to that of other, learned, naturalists. All in all, fishermen and fishmongers provided a

161 '[…] und ich alles selbst in meiner Hand gehabt, dieselbige nach dem leben abmahlen laszen, und wird ein jdes bey Seinem Nahmen genännet, und so viel ich bey einem jeden gelernt, in Seiner Natur, Kurtzlich ausz eigener erfahrung daszelbe beschrieben.' Add MS 6485, f3v.

162 'Und so mir Einer disze meine einfältige und geringe Arbeit, besser Verstehet, der wolle mirh, wo etwas gefehlt zu guth halten, Dann es von einem geringen Fischer und Weydman herkommet.' Add MS 6485, ff3v–4r.

163 'So hab ich im Nahmen desz Herrn mein Netz und Fischerkarn ausz geworffen, und ein wenig von dem was ich erlernt, und in Dreysig Jahren dabey Studiert hab, ein wenig wollen anzeigen.' Add MS 6485, f4r.

wide range of observations that ended up in the *Historia piscium*. Willughby and Ray qualified the observers that came from outside their own ranks either as ancient and most experienced and therefore trustworthy, or as possessing commendable curiosity despite lacking proper education.

4 Conclusion

Let us return to the *Historia piscium*'s discussions of the curious behaviours of the salmon one last time. A few lines after its peculiar matter of jumping is discussed, Ray addressed its mysterious eating habits:

> What food salmons use, because I see that authors disagree [on the matter], has to be consulted by experience.[164]

The matter of the salmon's diet had been discussed at a meeting of the Royal Society in 1678, where it was brought forth that fishmongers never found anything in the maws of salmon and that an (unnamed) lady, "very inquisitive in that kind", had observed the same.[165] The previous year, Johnson had written to Ray on the same subject. "I wonder as much that Fishers have not certainly determined whether Salmons live upon anything save Water, and what?"[166] He continued by noting that:

> I think only the Anglers have made the Observation of finding their Stomachs always empty; but I am persuaded that, if the Net-fishers would open any considerable Number, they would find in them Food indigested, which they seldom do, but sell them whole. Perhaps I may give farther Answer to this *Quaere*, and some others about *Whitsontide*; at which Time I purpose to go to our Coasts, and gather what I can.[167]

These discourses are indicative of the sorts of questions on which the Fellows pondered, and where they expected to find answers.

---

164 'Quo cibo utantur Salmones cum Autores diffentire videam, experientia consulenda est.' *Hist. pisc.*, 192.

165 Birch, *History of the Royal Society*, vol. 3, 425. See also Felicity Henderson, "Translation in the Circle of Robert Hooke," in *Translating Early Modern Science*, eds. Sietske Fransen, Niall Hodson, and Karl A.E. Enenkel (Leiden: Brill, 2017), 17.

166 Johnson to Ray, 16 April 1677 (OS), *Correspondence of John Ray*, 128.

167 Ibid.

The variety of places where Johnson suggests answers can be gathered fit well into recent widened conceptions on the part of historians with regard to the spaces where early moderns created (or perhaps stumbled upon) natural knowledge.[168] In London, fertile sites for assembling knowledge about fish encompassed – besides the rooms of Gresham College – coffeehouses, taverns, ports, fish markets, and the banks of the Thames.[169] Beyond the confines of the city, such locations included the coast of Cornwall and the (fish) markets of continental Europe. Each of these places allowed for the making of first-hand observations, but, even more importantly, for meeting those people whose observations of fish were informed by years of practice. These might be fishmongers, anglers, and net fishers. This chapter has tried to reconstruct the conversations between fishmongers, fishermen and Fellows so as to better apprehend what they actually consisted, and to analyse how these contributed to a deepened understanding of fish, whether individual species or as a whole. It has also emphasised how the extent and nature of these contributions might differ from person to person, relative to experience and skill. Taken together, the various examples discussed here demonstrate that exchanges with fishmongers and fishermen were not incidental, but rather were central to Willughby and Ray's project.

As this chapter has shown, the interactions between practical, artisanal communities and Fellows could be rather complicated. Fishermen and Fellows sometimes talked at cross-purposes, reminding us of similar difficulties in communication that arose in the Society's history of trades project.[170] Another issue was that while the Fellows appropriated knowledge from practical men and women for their discussions or publications, the practitioners themselves often were hidden well out of sight.[171] This also held true for other categories of fish connoisseurs, not discussed in this chapter, whose observations of fish

168 This historiography has become too vast to list exhaustively, but see, for example: Jim Bennett and Rebekah Higgitt, eds., *London 1600–1800: Communities of Natural Knowledge and Artificial Practice*, special issue of *British Journal for the History of Science* 52, no. 2 (2019): 183–343; Harkness, *The Jewel House*.

169 On the port of London as (continued) source for faraway species, see: Arthur MacGregor, "Patrons and Collectors: Contributors of Zoological Subjects to the Works of George Edwards (1694–1773)," *Journal for the History of Collections* 25, no. 1 (2013): 36.

170 Ochs, "The Royal Society of London's History of Trades Programme," 130.

171 Jasmine Kilburn-Toppin, "'A Place of Great Trust to be Supplied by Men of Skill and Integrity': Assayers and Knowledge Cultures in Late Sixteenth- and Seventeenth-Century London," in Bennett and Higgitt, *Communities of Natural Knowledge and Artificial Practice*, 222.

were drawn upon for the *Historia piscium* and which merit further consideration. Anglers, for example, also knew their way around fish. Willughby and Ray consulted Leonard Mascall's (*d.* 1589) well-known angling manual, *A Booke of Fishing with Hooke & Line, and of All Other Instruments There-unto Belonging* (London, 1590) when discussing the fact that while the carp was a relatively recent introduction to the waterways of England, it was now plentiful in rivers and ponds.[172] Anglers were also aware of whether a species was common or rare, and, as Johnson implied, knew what was in a fish's stomach. Other specific knowledge of fish pertained to their consumption. As we saw, *Historia piscium* offered glimpses of fish salters and cooks; and on occasion the taste and preparation of particular species of fish received attention in this book.[173]

For Willughby, Ray and other Fellows of the Royal Society, the value of interacting with fishermen and fishmongers lay in their repeated engagement with a large quantity and wide variety of fresh fish in an either living or recently deceased state. They did not only supply raw material, but also offered information that was crucial for the central tenet of the *Historia piscium*: to distinguish one species from the other by delineating their differences. Fishermen and fishmongers did not only know how to catch fish and how to tell them apart from another, but also commented on particular behaviours of certain species. It was on the basis of this sustained experience that Fellows regarded them as authorities in the world of fish. While, ultimately, the Fellows positioned themselves as prime arbiters on what passed as a credible observation and who qualified as a credible observer, this chapter has shown that they gladly ventured beyond the realm of the learned when seeking reliable and recognised authorities on fish.

As we saw in both the previous and current chapter, the *Historia piscium* was an attempt to create a universal work on the natural history of fish based on clearly defined principles, so that the proper relations between species and their names could be re-established and order restored in the wonderfully varied world of fish. Ray and Willughby's attempts to forge a new method for the study of fish were part of their broader aspiration to reform the study of nature. This ambitious agenda demanded that naturalists should privilege empirical examination of the physical characteristics of plants and animals over the claims of ancient or even more recent authorities. In this, they had to contend with all kinds of practical constraints, such as the at times imperfect

172 *Hist. pisc.*, 246 and Leonard Mascall, *A Booke of Fishing with Hooke & Line, and of All Other Instruments There-unto Belonging* (London: John Wolfe, 1590), 8.

173 See, for example: *Hist. pisc.*, 219, 320.

evidence that drawings or preserved specimens might present. This is why they sought out first-hand observations from a wide range of collocutors. The resulting work made it clear that the world of fish was well worthy of inquiry and yet still fundamentally difficult to fix in place. The dizzying variety of species, and the heuristic challenges that the study of them posed to the naturalist, required a further, even firmer grip on the order of fish. As we will see in the next chapter, Peter Artedi sought to accomplish precisely that. He developed an 'ichthyology' that drew up new demarcations not only between fish, but also between those who handled and studied them.

CHAPTER 3

Demarcating a Discipline: Peter Artedi's *Ichthyologia* and the Classification of Knowledge

It would have been one of the more curious specimens to end up on the dissection table of the anatomical theatre of professor Pieter Pauw (1564–1617) in the university town of Leiden. As the story goes, Joannes de Laet (1581–1649) brought a siren to Pauw, having witnessed the marine human being captured near the coast of Brazil while he served there as director to the Dutch West India Company. After it had been dissected, one hand and one rib of the specimen were given to Danish anatomist Thomas Bartholin (1616–1680), who proudly displayed them in his cabinet in Copenhagen. It is in his *Historiarum anatomicarum rariorium* (Amsterdam, 1654), a compilation of unusual anatomical cases, that we learn of this dissection.[1]

For the young and promising Swedish naturalist Peter Artedi, this report presented a problem. The question of whether beings like marine humans existed, and how they fitted into the larger scheme of Creation if they did, had long been a matter of contention.[2] Merman or mermaid parts were rare and coveted collectables – as we saw in Chapter 2, the Royal Society held the rib of a merman in its Repository.[3] The people involved in sightings of marine humans, such as Pauw, De Laet and Bartholin, were often reputable scholars. What to do? Artedi decided to include, in his *Ichthyologia, sive opera omnia de piscibus* [Ichthyology, or complete works about fish] posthumously published by his friend Carl Linnaeus in Leiden in 1738, the siren as a genus belonging to

1 Thomas Bartholin, *Historiarum anatomicarum rariorium* (Amsterdam: Ioannem Henrici, 1654), 169. See also: Christina Brito, *Humans and Aquatic Animals in Early Modern America and Africa* (Amsterdam: Amsterdam University Press, 2023), 123 and Peter Mason, *Infelicities: Representations of the Exotic* (Baltimore: Johns Hopkins University Press), 81. Unfortunately, the dates in this story do not appear to add up: Pieter Pauw passed away well before the WIC was founded in 1621.

2 A standard work on sirens in early modern Europe is Bernd Roling, *Drachen und Sirenen: Die Aufarbeitung und Abwicklung der Mythologie an den europäischen Universitäten* (Leiden: Brill, 2010).

3 See: Marjorie Swann, *Curiosities and Texts: The Culture of Collecting in Early Modern England* (Philadelphia: University of Pennsylvania Press, 2001), 28; Tara E. Pedersen, *Mermaids and the Production of Knowledge in Early Modern England* (Farnham: Ashgate, 2015), 51; Eric Jorink, *Reading the Book of Nature in the Dutch Golden Age, 1575–1715* (Leiden: Brill, 2010), 295, 298, 310–311.

 | DOI:10.1163/9789004549500_005

his order of cetaceans, or whale-like fishes.[4] He subjected it to the same natural historical principles as other fishes, but the uncertainty surrounding its existence put him in a delicate position. It led him to exclaim: "if only there were a true Ichthyologist, that could examine this animal, and find whether it is a fable, or a true fish? Not having seen the matter for myself, I prefer to not form a judgment, rather than to make any bold claim."[5] A 'true ichthyologist', then, as the proper expert on fishy matters like these, could, given the opportunity to examine a specimen in the flesh, settle the matter.

Artedi had become interested in the study of fish at a young age. He had been trying to match his own observations of fish to species descriptions published in books, but they had proved too vague and ultimately insufficient. That changed when he encountered a copy of the *Historia piscium*, the history of fishes by Willughby (regarding him as the sole author of the work), who he found "stood out above all others in his descriptions of species."[6] Artedi nevertheless believed the work lacked a firm grasp on the morphological relations between species of fish on different taxonomical levels. As he explains in the preface to the *Ichthyologia*:

> I then noticed that by not one of the Ichthyologists up to that time the distinct genera, nor their characteristics, nor their Species had been established; hereupon I began to examine all parts of Fish with great effort, so that I would see which parts of these were most similar according to Number, Figure, and Position, and which would be the most dissimilar, especially in regard to the fish which are consistent in their external appearance, from which the *Characteristics of Genera* and the *Genera* themselves are born.[7]

4 The whales continued to be counted among the fish. Artedi postulated that while most fishes respired through gills, some species of fish could also breathe through lungs. Linnaeus later ruled that cetaceans were mammals and not fish: see Carl Linnaeus, *Systema naturae*, ed. 10 (Stockholm: Lars Salvi, 1758), 17.

5 "Utinam existeret verus Ichthyologus, hoc qui examinaret animal, fabula utrum sit, an verus piscus? De re non visa potius est non judiciare, quam audacter quid pronuntiare." *Icht.*, *Genera piscium*, 81.

6 "[…] in descriptionibus specierum omnes supererat." Peter Artedi, *Ichthyologia, sive opera omnia de piscibus* (Leiden: Coenraad Wishoff, 1738), *Praefatio authoris*, sig. *r. The *Ichthyologia* will henceforth be referred to as *Icht.*, followed by the title of the part of the book from which the citation derives.

7 "Deinde distincta genera, eorumque characteres & Species à nullo Ichthyologorum adhucdum constituta esse animadvertabam; hinc magno nisu omes Piscium partes examinere incipiebam, ut viderem quaenam illorum partes secundum Numerum, Figuram & situm maxime convenirent, & quaenam maxime discreparent, imprimis in Piscibus facie externa

Despite its modest octavo format and lack of any illustrations, the *Ichthyologia* was a far from unassuming book. In it, Artedi presented an elaborate system for the classification of fishes, introducing the taxonomical ranks of class, order, and genus, and offering clear definitions of both species and varieties. As we saw in the previous chapters, naturalists had grouped species together based on their habitat, their letter of the alphabet, or according to their morphological characters or physiological features. Artedi clearly and distinctly allocated species into ranks based on shared characteristics, which could consist of intricate combinations of external features and inner parts. His aim was to uncover the intricate, taxonomical arrangements of fishes through the application of uniform and consistent rules for their description, naming and classification. In so doing, he sought to impose a unity, consistency and logical order onto the natural history of fish that he found to be sorely lacking.

Artedi's system was an expression of the widely shared search for regularities and patterns in nature. Three years before the *Ichthyologia* appeared, Linnaeus had published the first edition of his *Systema naturae* (Leiden, 1735) in which he unfolded his classification system for minerals, plants, and animals. In the case of fish, he adopted Artedi's method of classification.[8] Both Linnaeus' and Artedi's works fitted well into the broader development in eighteenth-century Europe that saw the quantification of nature, of measuring and calculating, ordering and systematising it.[9] This entailed turning observations that were rich with detail and tied to a specific time and place, such as medical cases or weather diaries, into synthetic, general observations on the changing character of a disease or deviations of an average climate.[10] Observations of nature were compressed into tables from which correlations might be deduced.[11] Analogous to this development, naturalists like Artedi and Linnaeus increasingly set their sights on arriving at uniform and consistent principles with which they could name, describe and classify minerals, plants

convenientibus, unde *Characteres Generum* & ipsa *Genera* nata sunt [...]" *Icht., Praefatio authoris*, sig. *r.

8 Broberg, *Carl Linnaeus*, 142.

9 John Heilbron, "Introduction," in *The Quantifying Spirit in the Eighteenth Century*, eds. Tore Frängsmyr, J.L. Heilbron, and Robin E. Rider (Berkeley: University of California Press, 1990), 2.

10 J. Andrew Mendelsohn, "The World on a Page: Making a General Observation in the Eighteenth Century," in Daston and Lunbeck, *Histories of Scientific Observation*, 69–89.

11 Lorraine Daston, "Super-Vision: Weather Watching and Table Reading in the Early Modern Royal Society and Académie Royale des Sciences," *Huntington Library Quarterly* 78, no. 2 (2015): 189.

and animals.[12] In formulating these principles, numbers and measurements played a role to such a considerable extent that historians have characterised such pursuits as mathematical, arithmetical, or geometrical.[13] But how could one quantify a fish? Plants and animals were not that easily condensed in formulae, precisely because of the sheer variety of parts, shapes and colours they exhibited. While categorisations of nature based on morphological features had been around for quite some time, as we have seen in the previous chapters, over the course of the eighteenth century naturalists began to place strong emphasis on the enumeration of characteristics that they considered consequential, such as the pistils of flowers or, in the case of Artedi, the fin rays of fish. Artedi's ambition to design a comprehensive system to classify nature was thus not, in itself, unique. The seventeenth and eighteenth centuries in particular saw a proliferation of competing models aimed at classifying animals based on their physiologies. The one that Artedi devised for fish proved to be particularly successful however, and naturalists would apply his system in their examinations of fish for decades to come. Bloch, the subject of the next chapter, was one of them.

The story of the siren highlights some of the core questions of this chapter. How can one arrive at true, certain knowledge about fish? How can all species be fitted into one system? Who gets to decide, and on the basis of what? Who was a true ichthyologist? This chapter explores these questions through the *Ichthyologia*. In this book, we find the first explicit articulation of what 'ichthyology' is (or, at least, what Artedi thought it ought to be) as well of the very first definition of its prime practitioner: the 'ichthyologist'. Besides seeking to understand what this term meant according to Artedi, it is worthwhile considering how he wielded it to lend credence to his own system. This entails being attentive to the context in which the term 'ichthyology' emerged to denote a specialised knowledge on fish, a process that remains seldom reflected upon by historians of natural history.

The chapter will first introduce Peter Artedi, and shed light on his attempts to establish his name as a naturalist. It will then look more closely into his *Ichthyologia* and analyse its structure and the rhetorical strategies it employs. Subsequently, it will examine Artedi's classification system, *viz.* his division into classes, orders, genera, and species. Lastly, it will reflect on what impact

12 John E. Lesch, "Systematics and the Geometrical Spirit," in Frängsmyr, Heilbron and Rider, *The Quantifying Spirit*, 75.

13 Heilbron, "Introduction," 20–21; Lesch, "Systematics and the Geometrical Spirit," 73–111; William T. Stearn, "Carl Linnaeus and the Theory and Practice of Horticulture," *Taxon* 25, no. 1 (1976): 24.

the publication of this system had on what was considered to be of import to learned inquiries into fish, and how, by privileging certain characteristics of fish over other ones, it prescribed which aspects of the piscine population should be studied, preserved, and presented.

1 The Short Career of Peter Artedi

"One of the big mysteries in Swedish history of science" is how Gunnar Broberg, in his biography of Linnaeus, characterised Artedi.[14] A biographer of Artedi does indeed confront a challenging lack of source material. The naturalist died young and only one of his works – the *Ichthyologia* – appeared in print, and that posthumously. And although this work contained, as we will see, a daringly innovative system for organising living nature, it has not attracted sustained attention beyond that of specialists. Of course, the fact that up until recently the work was only available in Latin has not helped to make his writings accessible to a larger audience, either scholarly and beyond. That is changing, thanks to the scrupulous efforts of Theodore W. Pietsch and Hans Aili to translate both Artedi's published and unpublished writings into English while also placing his ideas into a larger context of debates around classification.[15]

A lot of what is known about Artedi comes from the biographical note that was written by his much better-known friend and collaborator Linnaeus, and which opens the *Ichthyologia*. This account has formed the basis for many of the later biographies of Artedi.[16] It is an illustration of how Artedi's name has become almost inextricably connected with that of Linnaeus, and how his life's

14 Broberg, *Carl Linnaeus*, 74.

15 Hans Aili and Theodore W. Pietsch, *Peter Artedi: Reformer of 18th Century Zoology*, vol. 1: *Peter Artedi's Life and Works* (Stockholm: Stockholm University Press, 2024).

16 The first extensive biographical sketch is that of Einar Lönnberg, *Peter Artedi: A Bicentenary Memoir Written on Behalf of the Swedish Royal Academy of Science*, trans. W.E. Harlock (Uppsala: Almqvist & Wiksells, 1905). More recent publications are, in chronological order, Daniel Merriman, "Peter Artedi – Systematist and Ichthyologist," *Copeia* 25, no. 1 (1938): 33–39; Gunnar Broberg, "Petrus Artedi in his Swedish Context," in *Proceedings of the Fifth Congress on European Ichthyology*, eds. Sven O. Kullander and Bo Fernholm (Stockholm: Swedish Museum of Natural History, 1987): 11–15; and Alwhyne Wheeler, "Petrus Artedi, Founder of Modern Ichthyology," in Kullander and Fernholm, *Proceedings of the Fifth Congress on European Ichthyology*, 3–10; Theodore W. Pietsch and Hans Aili, "Swedish Naturalist Peter Artedi (1705–1735) and His Place in the History of Biosystematics as Exemplified by his *Ichthyologia sive opera omnia piscibus* of 1738." *Zootaxa* 5169, no. 6 (2022): 589–598 and Aili and Pietsch, *Peter Artedi: Reformer of 18th Century Zoology*, vol. 1, 15–19, 25–41.

story has been mediated by the latter's pen. Conversely, few of the works that detail Linnaeus' life even mention Artedi's name.[17] The lack of source materials surrounding Artedi is especially marked when compared to those concerning Linnaeus. This contrast becomes clear at once when considering their respective correspondences that have come down to us: while only one letter by Artedi is currently known to remain, the Linnean Society of London alone retains several thousand letters sent by and to Linnaeus.[18] During his long life, furthermore, Linnaeus published a plethora of natural historical works and issued updates of some of these, like his influential *Systema naturae* (Leiden, 1735), of which he published the twelfth and last edition in 1766.

Due to his rigorous reform of the study of fish, Artedi has been called the 'father' or 'founder' of ichthyology.[19] On a monument that was unveiled in the Zoological Garden of Amsterdam in 1905, Artedi was crowned the 'Prince of Ichthyology', just as Linnaeus has often been dubbed the 'Prince of Botanists.'[20] Such honorary titles underline Artedi's importance for the field of ichthyology, and natural history more broadly. Yet how did he make this name for himself? In what ways did he present himself and his ideas to the learned world, even during the admittedly short span of his life? While Linnaeus has been the focus of historical inquiries that go well beyond a celebration of his accomplishments to in-depth studies of his ideas, practices and his keen sense for advancing his work – so much so that Linnaean studies has become a flourishing subset of the history of science – the ways in which Artedi navigated this learned landscape remain underexplored.[21]

By looking at this still rather unknown naturalist bent upon establishing a place for himself in the study of natural history, light can be thrown on the efficacy of various strategies for doing so. Artedi is particularly interesting because his name has become so intricately connected to the field of ichthyology, a process he actively tried to shape. This he did, of course, by designing

17 An exception being Lisbet Koerner [Rausing], "Linnaeus in His Time and Place," in Jardine, Secord and Spary, *Cultures of Natural History*, 150.

18 The Uppsala Universitetsbibliotek holds hundreds more, digitised as part of the *The Linnaean Correspondence* and accessible via http://www.alvin-portal.org (last accessed 15 February 2025).

19 Hans Aili and Theodore Pietsch, *Peter Artedi: Reformer of 18th Century Zoology*, vol. 1, 1. Daniel Merriman, "Peter Artedi – Systematist and Ichthyologist," *Copeia* 25, no. 1 (1938): 33; Alwhyne Wheeler, "Petrus Artedi, Founder of Modern Ichthyology," in Kullander and Fernholm, *Proceedings of the Fifth Congress on European Ichthyology*, 6.

20 "Ichthyologorum longe princeps", Kerbert, *Peter Artedi*, 15.

21 Hanna Hodacs, Kenneth Nyberg and Stéphane Van Damme, "Introduction: De-centring and Re-centring Linnaeus," in *Linnaeus, Natural History and the Circulation of Knowledge*, eds. Hodacs, Nyberg and Van Damme (Oxford: Voltaire Foundation, 2018), 9–15.

his classification system, but also by styling himself as a certain type of scholar particularly attuned to the study of fish: the ichthyologist, a position for which he was the first to write a job description, as we will see. This chapter, therefore, is less concerned with questions of primacy and priority and all the more with understanding how Artedi attempted to make his name synonymous with the field of ichthyology. A biographical note is now in order.

Artedi was born in 1705 in the Swedish province of Ångermanland, in a family of modest means. His father was a minister in the Anundsjö parish. When the family moved to the coastal province of Nordmaling in 1716, Artedi was sent to grammar school in Härnösand – where he spent his free hours collecting plants and dissecting fish.[22] As he was expected to follow in his father's footsteps and become a clergyman, he enrolled at the School of Divinity at Uppsala University in 1724 with the support of an annual stipend.[23] After two years, however, he switched to the faculty of medicine so that he could fully apply himself to the study of natural history.[24] He mostly worked alone, not being surrounded by students as eager as he. This changed when the young Linnaeus arrived at Uppsala university in 1728. Linnaeus, who came from a similar humble background, had transferred from Lund to take up the study of natural history. The students struck up a friendship based on their shared enthusiasm. For the next seven years, they examined nature in tandem. Three manuscripts that Artedi produced in the late 1720s are still extant: an overview of the herbs and trees of Nordmaling; a classification system for hirsute animals, and a catalogue of fish of the Baltic region.[25] Each of these works, to be discussed in more detail below, reveal his preoccupation with the categorisation of species. It was

---

22 Lönnberg, *Peter Artedi*, 5; Artedi's father had "been admonished by the bishop on account of his connection with an adventuress, Maja Stina Fröling, who had acquired both the parish funds and the little money he had [...]," see: Broberg, "Petrus Artedi in his Swedish Context," 11.

23 Ibid., 15.

24 Some of his lecture notes and a bound manuscript of book summaries remain. The letter is glued to the inside of the cover of the latter (communication via email with Thomas Artedi, descendant of the Artedi family, whose transcription I have used). Kungliga Biblioteket (hereafter KBS), Stockholm, MS X1002. I thank Anna Svensson for translating this letter into English.

25 Peter Artedi, *Kårt Förtekning på de Träen, Buskar åg Örter, såmm wäxa sponté wid Nordmalings Prästebord äller i de närmaste byar där åmmkring Äfter Dänn Alldra-simplaste åg Klaraste Methoden i årdning satte av Petro Arctædio A:o 1729 in Februario*. Special Collections Department of Uppsala Universitetsbibliotek (hereafter UUB), D 82 a; *Catalogus Piscium Maris Balthici ut et Fluviorum ac Lacuum Sveciae*, BL, Add MS 3870; and *Idea Institutionum Trichozoologiae*, Bergius Library, Universitetsbibliotek Stockholm (hereafter SUB), Stockholm, H.VII.8.1.n.4.

also around this time that he started preparing his extensive manuscript on the study of fish.[26] Linnaeus recounts how, on leaving Uppsala, they promised one another that should either of them die, the survivor would see to it that their late friend's observations and investigations would be published.[27]

Linnaeus' and Artedi's approaches to natural history contain many similarities, which is not surprising. After all, they spent the larger part of their student days (from the late 1720s to early 1730s) in close contact. We can infer from Artedi's manuscripts that he spent much of his time observing his immediate, natural surroundings, examining plants, quadrupeds and fish. Given that his species descriptions include some subsurface anatomical detail that extends beyond those features visible on the outside of the fish, we can conclude that he must have occasionally performed dissections. Linnaeus devoted his days to similar pursuits. As he explains, he had daily meetings with Artedi during which they disclosed whatever they had discovered.[28] Together, they discussed how they might bring order to nature's varied productions. Their collaboration also entailed competition. Linnaeus remarks how after a few years and some "[...] protracted labour to attain premier position in Ichthyology, I was finally fain to admit my inferiority to my rival, and thenceforth I left that subject entirely in his hands, as also the study of Amphibia, while, on the other hand, he willingly acknowledged that I was ahead of him in knowledge of Birds and Insects."[29] Linnaeus and Artedi thus each came to focus themselves on particular types of animals. The exception were the quadrupeds, on which they both continued their work. Artedi called these creatures *trichozoa* [hairy animals], Linnaeus named them *mammalia* [mammals] for their mammary glands; the latter term stuck.[30]

Artedi's zeal for classification is already evident in his earliest manuscript. In February 1729, he completed a flora of his native Nordmaling under the title *Kårt Förtekning på de Träen, Buskar åg Örter, såmm wäxa sponté wid Nordmalings Prästebord äller i de närmaste byar där åmmkring* [A Short

---

26 In the preface, dated 1735, Artedi explains that he had begun his examination of fishes eight years ago. *Icht.*, *Praefatio authoris*, n.p.

27 Lönnberg, *Peter Artedi*, 22.

28 Carl Linnaeus, "Vita authoris," in Artedi, *Ichthyologia*, sig. *4r.

29 "[...] certabat uterque nostrum de victoria obtinenda in Ichthyologicis, donec post diuturnos labores ei victas deberem dare manus, totum hinc studium hoc ipsi commisi, ut & amphibiorum notitiam; contra vero in Avium & Insectorum notitia ille libens mihi cedebat primas." Ibid., sig. *3v, adapted translation from Lönnberg, *Peter Artedi*, 11.

30 For their respective treatments of this category, see: Theodore W. Pietsch and Hans Aili, "Peter Artedi's 'Idea institutionum Trichozoologiae' and the Classification of Mammals." *Journal of Natural History* 57, no. 17–20 (2023): 1066–1079.

List of the Trees, Bushes and Herbs that are Indigenous to the Glebe-Lands in Nordmaling and the Villages Lying in its Immediate Vicinity].[31] The title went on to indicate that it was "put in order after the most simple and clear method";[32] something that, as we will see, was a recurrent theme for Artedi. His taxonomical organisation was inspired by the system of the French naturalist Joseph Pitton de Tournefort (1656–1708).[33] In his *Institutiones rei herbariae* (1700), de Tournefort had made a clear distinction between a genus and species in the realm of plants, grouping those plants together that resembled one another in flower and fruits.[34] Genera were collections of species that corresponded to one another in certain parts (for example, the number, shape and symmetry of the petals), and differed in these respects from the members of other genera.[35] Artedi came to consider the rank of genera as an integral part to the study of natural history because of its value for grouping species. In his undated manuscript entitled *Idea Institutionum Trichozoologiae* [Outline of principles of the zoology of hirsute animals] he proposed a classification scheme for mammals that subdivided them into genera.[36] He did the same for fish, as will be discussed in more detail later in this chapter.

Artedi and Linnaeus came into contact with contemporary ideas in natural history through their university's library, rather than through lectures. Linnaeus complained about the lack of instruction in the field of natural history at their university, saying that: "I myself never had the opportunity of attending a single lecture on Botany, either private or public."[37] After the two students had finished their degree, and their funding had run out, they sought to leave Uppsala and try their hand elsewhere. In order to effect this, they applied for stipends with the Royal Society of Letters and Sciences in Uppsala (known as the *Societas regia literaria et scientarium*). This society, modelled after the

31 As cited in note 16.

32 "*Äfter Dänn Alldra-simplaste åg Klaraste Methoden i årdning satte* […]" UUB, D 82 a, title page.

33 Artedi probably took hold of de Tournefort's work through Lars Roberg (1664–1742), who was professor of medicine at Uppsala University. A catalogue of his library, which can be found in UUB, D60, shows he possessed at least one of de Tournefort's botanical books.

34 Yves Cambefort, "How General are Genera? The Genus in Systematic Zoology," in *The Oxford Handbook of Generality in Mathematics and the Sciences*, eds. Karine Chemla, Renaud Chorlay and David Rabouin (Oxford: Oxford University Press, 2016), 260.

35 This definition comes from Merriman, "Peter Artedi – Systematist and Ichthyologist," 38–39.

36 The contents of this manuscript have been transcribed in Orvar Nybelin, "Tvenne Opublicerade Artedi-Manuskript," *Svenska Linnésällskapets Årsskrift* 18 (1935): 58–77. See also Ali and Pietsch, *Peter Artedi: Reformer of 18th Century Zoology*, vol. 1, 112–114.

37 Lönnberg, *Peter Artedi*, 8.

Royal Society in London, had been founded in the early eighteenth century.[38] Linnaeus subsequently received the support of the Uppsalian Society for his natural historical expedition to the province of Lapland in 1732.[39] Artedi seems to have applied for another funding stream offered by this academy, the 'Stipendium Stiglerianum' set up by the merchant Jacob Stiegler (1649–1716) in May 1734; probably without success, for he later appealed to his inlaws Peter Biur and Jonas Liungberg for funding.[40] One still needed to be prosperous to be a naturalist.

Artedi's entreaty to his inlaws was successful. A decade after enrolling at Uppsala University, he left Sweden to seek out new flora, new fauna and, most pressingly, new opportunities. In September 1734, he set sail for England.[41] As the previous chapters has shown, this was a suitable destination for someone interested in natural history. Artedi had prepared this trip well. Besides securing financial support from his inlaws, he also capitalised on his connections in another way. He found Jacob Serenius (1700–1776), chaplain to the Swedish congregation in London, willing to write him a letter of recommendation to Hans Sloane, who had by this time become the president of the Royal Society. In this letter, Serenius stated that he thought that Sloane would be "pleased to grant him [Artedi] when you find his skill in ictyology [*sic*] and other parts of natural history."[42] When Artedi visited Sloane in London, he brought along a gift: his manuscript entitled *Catalogus Piscium Maris Balthici ut et Fluviorum ac Lacuum Sveciae* [Catalogue of Fish in the Baltic Sea and the Rivers and Lakes of Sweden]. It was a result of his prolonged study of fishes in his vicinity; the last few pages gave a brief summary of the *Ichthyologia*, which he probably intended to publish in the foreseeable future.[43] Artedi presented himself,

38 Tore Frängsmyr, "Linnaeus in his Swedish Context," in *Contemporary Perspectives on Linnaeus*, ed. John Weinstock (Lanham: University Press of America, 1985), 183–186.

39 For the background of this enterprise, see: Lisbet Koerner [Rausing], *Linnaeus: Nature and Nation* (Cambridge, Mass.: Harvard University Press, 1999), 56–81.

40 Application, dated 22 May 1734; UUB, Waller Ms se-00136. This stipend is described in *Samling af testamenten och författningar om stipendien* (Uppsala: Johan Fr. Edman, 1795), 62–78.

41 Merriman, "Peter Artedi," 35.

42 Jacob Serenius to Hans Sloane, 26 October 1734 (OS) as reproduced in Orvar Nybelin, "Kring Petrus Artedi's vistelse i England 1734–1735," *Svenska Linnésällskapets Årsskrift* 49 (1966): 23–25.

43 BL, Add MS 3870. The contents of this manuscript have been reproduced in Nybelin, "Tvenne Opublicerade Artedi-Manuskript," 78–90. For an English translation, see: Theodore W. Pietsch and Hans Aili, "Peter Artedi's Catalogue of the Fishes of the Baltic Sea: An English Translation with an Introduction and Commentary," *Zoological Journal of the Linnean Society* 189, no. 3 (2020): 975–997.

and was presented by others, as a naturalist particularly skilled in the study of fishes. As broad as his natural historical interests ranged, he aimed to make his name as an ichthyologist.

As a centre of global commerce, a city like London was a hub for the study of specimens from faraway places. Artedi encountered animals that he had never seen before, and he examined them with curiosity and eye for detail. An appendix to the *Idea Institutionum Trichozoologiae* includes his notes on a "panther from Buenos Aires" that he observed in the city alongside a leopard, as well as tigers and a hyena sighted in Moorfields, an area for markets and exhibitions on the northern edge of the city.[44] Luckily, there were also plenty of fish to be found. The species descriptions in the *Ichthyologia* make it possible to retrace some of Artedi's itinerary in London. He made use of the opportunity to study Sloane's expansive collection by describing some of its species of boxfish.[45] Sloane was also the patron of James Salter (d.1728), who preferred to call himself Don Saltero, and owned a Chelsea coffee house in which he displayed a broad range of *artificialia* and *naturalia*.[46] In this coffee house, Artedi further saw species of boxfish.[47] He took note of the whale that found its way into the city in November 1734.[48] Artedi also visited the Swedish publican Lars Lilja (d.1744) at his establishment in Shadwell, as well as taverns such as the Nag's Head, Green Dragon and White Bear.[49] These pubs were frequented by sailors returning from a journey to faraway places, and Ovar Nybelin has conjectured that they may have used dried exemplars of fish acquired during these travels to pay their drinking bills – a hypothesis that does correspond to the value that was attributed to rare natural historical objects.[50] Artedi's visits to collections in London, and later Amsterdam, led him to remark that the fish he had seen were "altogether rare and curious, so I have seen more curiosities in this part of natural history than someone who has travelled through all of Europe."[51]

44 Theodore W. Pietsch and Hans Aili, "Peter Artedi's Early Observations of the Spotted Hyena and Other Exotic Animals During a Visit to London (1734–1735)," *Archives of Natural History* 50, no. 2 (2023): 410–416. The authors raise doubt on the identification of the panther.

45 "vidi in museo Hans Sloane", "vidi apud Sir Hans Sloane", *Icht., Genera piscium*, 56–57.

46 For an idea of the contents of this collection, see: James Salter, *A Catalogue of the Rarities, to be Seen at Don Saltero's Coffee House*, vol. 10 (London: s.n., 1731).

47 *Icht., Genera piscium*, 58.

48 "Ex observatione propria, Londini 1734 in Novembri", *Icht., Genera piscium*, 77 and "visa a me Londini anno 1734" and *Icht., Descriptiones specierum*, 107.

49 Nybelin, "Kring Petrus Artedi's vistelse i England 1734–1735," 27.

50 Ibid., 10–12.

51 "[…] allesammans rare, och curieuse, så att jag sedt flera curieusiteter utj den delen af Hist. naturali, än den som rest genom hela Europa." KBS, MS X1002.

After a brief, albeit expensive, stay in London Artedi travelled to the Dutch Republic in the summer of 1735, with the hope of attaining a doctorate in medicine.[52] He had, however, used up all his financial reserves and could not afford to enrol straightaway. By chance, he was reunited with Linnaeus in Leiden. Linnaeus had received his doctorate at the University of Harderwijk a month previously. He had also been commissioned by the wealthy merchant banker George Clifford (1685–1760), to curate the garden of his estate Hartekamp, near the village of Heemstede. Upon learning of Artedi's financial struggles, Linnaeus introduced him to the wealthy apothecary and collector Albertus Seba (1665–1736), who commissioned Artedi to describe the fish in his collection.

Linnaeus' introduction to Albertus Seba must have appeared propitious. Seba was a renowned collector of *naturalia* from the East and West Indies and beyond – tsar Peter the Great (1672–1725) had visited his collection in Amsterdam and purchased it for an enormous sum in 1716.[53] Seba had meanwhile built another impressive collection of natural curios, which he was codifying into print under the abbreviated title *Thesaurus* [Treasure].[54] In the one letter of Artedi's that has survived, written to his relatives a few days before his death in September 1735 in Amsterdam, Artedi narrates that: "[t]he first time I was in Amsterdam, there was much talk of an Apothecary, that will soon publish a great work of natural history, I visited him, and when he noticed that I was at home in Ichthyology, he convinced me to stay on a while in Amsterdam and describe his East-Indian and American fish, for he does not know any of them."[55] Artedi's intimate knowledge of the natural history of fish, which he explicitly refers to as ichthyology, made him a suitable candidate for this task.

Artedi assumed this task on the understanding that his name would be included in the work as an author.[56] It is likely that he had seen the book's

52 Merriman, "Peter Artedi," 35.

53 Margócsy, *Commercial Visions*, 89. For details of the sale, see: Jozien J. Driessen-van het Reve, *De* Kunstkamera *van Peter de Grote: De Hollandse Inbreng, Gereconstrueerd uit Brieven van Albert Seba en Johann Daniel Schumacher, uit de Jaren 1711–1752* (Hilversum: Verloren, 2006), esp. 107–117.

54 Albertus Seba, *Locupletissimi rerum naturalium thesauri accurate description et iconibus artificiosissimis expressio* (Amsterdam: apud Wetstenium, & Gul. Smith, & Janssonio-Waesbergios, 1734–1765).

55 "När jag första gången kåm hit till Amsterdam taltes mycket åm en Apothecare, såm härstädes gier ut ett stort werck utj historia Naturali, jag hälsade på hånom, och effter han märchte jag war hemma utj Ichthyologien, så öwertalte han mig att bliwa en tid qwar i Amsterdam och gifa en beskrifning på hans Ost-Indiska och Amerikansca fiskar, ty han kiänner sielf icke en enda." KBS, MS X1002.

56 Ibid.

first volume, which was published in 1734. The second volume had either just appeared or was close to doing so. In his letter, Artedi stated to have heard that Her Majesty of Sweden, Queen Ulrika Eleonora (1688–1741), had bought these splendidly illustrated volumes for the library of Drottningholm Palace. Being a recognised contributor to Seba's project would help further Artedi's reputation. Possibilities like these seemed to have been something of a preoccupation for him, as he continued to write: "[i]t is a hard thing when one first begins to show oneself in the world and one's name begins to be known, but can't fulfil his ambition and rise as far as he would wish."[57] He was glad to have met Seba because the commission he received meant that he could exert himself and get something for his trouble. The manuscript containing his notes still remains, and shows how he carefully described and classified these species. It has the title *Manuscriptum ichthyologicum quod Petrus Artedi elaboravit in usum Thesauri Seba* [Ichthyological manuscript compiled by Peter Artedi for the use of Seba's *Thesaurus*].[58] Just like Artedi's other manuscripts, it offers species descriptions listing the main characteristics of each species ordered according to their genus; he used some of these descriptions in the *Ichthyologia*.[59]

While Artedi's name was indeed mentioned in the preface to the third volume of the *Thesaurus*, which was entirely devoted to aquatic flora and fauna, he would never see it in print (nor, incidentally, would Seba – the book was finally published in 1759).[60] Before he could complete the task, Artedi drowned

57 "Det är en swår ting när man först begynner wisa sig i werlden och ens namn begynner bli bekant att man då intet kan hålla fort och bringa det så högt såm man vill, hwilket åffta obigerar en att han icke wåndar att bliwa det han kan." Ibid.

58 Peter Artedi, *Manuscriptum ichthyologicum quod Petrus Artedi elaboravit in usum Thesauri Sebani*, Library of Congress (hereafter LC), Washington D.C., QL618.15 A78 1735; it is mentioned in Pietsch and Aili, "Peter Artedi's Catalogue," 3, and described in more detail in Peter Merriman, "A Rare Manuscript Adding to our Knowledge of the Work of Peter Artedi," *Copeia* 2, no. 1 (1941): 64–69. This is the manuscript Margócsy writes having looked for in vain in *Commercial Visions*, 243. It has since resurfaced, see: Theodore W. Pietsch and Hans Aili, "Peter Artedi's 'Manuscriptum ichthyologicum', A Source for Albertus Seba's Locupletissimi rerum naturalium thesauri accurata descriptio (1759)." *Archives of Natural History* 50, no. 1 (2023): 118–132.

59 Its title page states it is "an ichthyological treatise describing more than a hundred of the principal fishes of Amboina and Surinam, which are arranged according to the natural method, with their distinct genera and new specific differences, and with the principal synonyms used by authors in previous descriptions"; translation taken from Merriman, "A Rare Manuscript," 66–67.

60 Artedi's descriptions were edited by Arnout Vosmaer, who would later be appointed director of the menagerie of Stadholder William v in The Hague. See: Margócsy, *Commercial Visions*, 98, 105–106; L.B. Holthuis, "Albertus Seba's 'Locupletissimi rerum

in an Amsterdam canal following an overly convivial evening at Seba's house in 1735.[61] Now that his friend's life had been brought to an abrupt halt, it was left to Linnaeus to prepare his nearly finished manuscript on fish in print. Although Seba initially was unwilling to release the manuscript,[62] Linnaeus managed to secure it for a little less than 100 guilders with the financial support of his patron Clifford.[63] He finally published the *Ichthyologia* in 1738. It appeared in Latin, the scholarly language that Artedi had written it in.[64]

Artedi died before his efforts to establish a name for himself as a naturalist had borne fruit. Linnaeus, by way of contrast, would have ample chance to make his name, and he took full advantage. Through the patronage of various wealthy individuals and the awarding of various stipends, Linnaeus published a plethora of works which succeeded in getting both his name and his ideas out onto the stage of the natural historical world. In 1741, he was appointed professor in Medicine, with special emphasis on botany, at the University of Uppsala. He became rector in 1750, and was ennobled in 1761, all the while continuing his research.[65]

The friendship between the two naturalists, and its unfortunate end, offers a striking parallel to that of Willughby and Ray. Linnaeus himself was aware of this connection, likening his collaboration with Artedi to that of the English naturalists.[66] Even though Artedi's life was cut tragically short, the career trajectories both he and Linnaeus traversed illustrate some of the hardships that aspiring naturalists encountered when building a name for themselves. Salaried positions in the field of natural history remained few and far between, so that the ability to dedicate one's life towards investigating nature often depended on one's financial resources, or those of one's family. Even from the little remaining material pertaining to Artedi, it is possible to gain a better

naturalium thesauri ...' (1734–1765) and the 'Planches de Seba' (1827–1831)," *Zoologische Mededelingen* 43, no. 19 (1969): 243, 247.

61 The possibility that this unfortunate event might not have been, in fact, an accident is explored in a murder mystery by Theodore W. Pietsch, *The Curious Death of Peter Artedi: A Mystery in the History of Science* (New York: Scott & Nix, 2010).

62 Seba appeared to have taken the manuscript as collateral for unpaid bills. Broberg, *Carl Linnaeus*, 135.

63 Ibid., 136.

64 A Swedish translation of this work was delivered by Hans Aili, *Ichtyologia. Det vill säga alla verk om fiskarna*, ed. Jakob Christensson (Stockholm: Kungliga Skogs- och Lantbruksakademien, 2022).

65 For the honours that Linnaeus accrued towards the end of his life, see: Broberg, *Carl Linnaeus*, 339–342.

66 Charmantier, Johnston and Smith, "The Legacies of Francis Willughby," in Birkhead, *Virtuoso by Nature*, 380.

understanding of how he tried to establish his name as a naturalist, or rather an ichthyologist. After undertaking his studies, he further developed his classificatory system and seized opportunities to demonstrate his command over the natural history of fish to influential naturalists such as Sloane and Seba in the hopes of gaining recognition. The remainder of this chapter will discuss how Artedi confidently presented a new approach for the natural historical study of fishes, and how he allocated himself an important role in this process.

2 Demarcating a Field

Artedi dedicated himself to finding patterns in living nature. He sought to lay bare the underlying structures of the world of fish one gill and one fin ray at a time. Out of these assiduous efforts, he developed a new approach for studying these creatures which entailed classifying them into a system of hierarchical tiers on the basis of physical features. This is the method he detailed in his book. Yet his work had a larger aim, namely to establish how this method formed the basis for the study of ichthyology, a subdiscipline of natural history that could – and should – be demarcated from the study of other classes of animals on the one hand, and more practical kinds of knowledge about fish on the other. We will now turn to how Artedi drew the lines of this new subdiscipline of natural history in his *Ichthyologia*.

The *Ichthyologia* was published in 5 volumes in 1738. Linnaeus had lived up to his promise by ensuring the work of his friend made it into print. He found a suitable publisher in Coenraad Wishoff (*c.*1690–1763), the Leiden bookseller with whom he had printed a few of his own works and who appears to have specialised in scholarly publications.[67] Wishoff advertised the work in the newspaper the *Leydse courant* alongside Linnaeus' *Classes plantarum* and a reissue of Johannes Swammerdam's 1667 *Tractatus physico-anatomico-medicus de respiratione usuque pulmonum*, revised by Albert von Haller (1708–1777) – some serious scholarly tomes.[68] The advertisement praised the *Ichthyologia* as an "opus sine pari, curante Linnaeo", an incomparable work that had been

67 Wishoff seems to have specialised in learned books: among his portfolio were works on medicine and natural history, as well as many academic dissertations and a mathematical dictionary. No extensive study on the Wishoff booksellers exists, but they are mentioned in André Bouwman, Ed van der Vlist, Berry Dongelmans and Paul Hoftijzer, eds., *Stad van Boeken: Handschrift en Druk in Leiden, 1260–2000* (Leiden: Primavera, 2008), 202, 223; Rietje van Vliet, *Elie Luzac (1721–1796): Boekverkoper van de Verlichting* (Nijmegen: Vantilt, 2005), 53.

68 Advertisement for books of Conrad Wishoff, *Leydse Courant*, May 7, 1738, verso.

attended to by Linnaeus.[69] No price was listed, but it must have been relatively affordable.[70]

At first sight it appears a quite unassuming book, especially when compared to the luscious, illustrated natural historical folios that were published in the eighteenth century, like Seba's *Thesaurus*.[71] Not only was the *Ichthyologia* published in smaller, cheaper octavo format, but not a single image was to be found amongst its five hundred or so pages. The book opens with the biographical sketch by Linnaeus, as said, followed by Artedi's own preface that expatiates on the genesis, aim and structure of his work. The book comprises five parts, each endowed with its own title page. The first, *Bibliotheca ichthyologia*, offers a bibliographic overview; the second, the *Philosophia*, a theoretical framework laid out in aphorisms; the third, *Genera piscium*, a discussion of genera; the fourth, *Synonyma*, an overview of synonyms; and the fifth and last, *Descriptiones specierum piscium*, species descriptions of fishes. The third part of the book is preceded by a pair of dedicatory letters from Linnaeus' hand. One is addressed to his maecenas Clifford, praising his financial support in printing this work, the other to Artedi's inlaws Liungberg and Biur, for having done the same.[72] The publication of this work gave Linnaeus the occasion to further strengthen these ties.[73]

Linnaeus had a hand in shaping the book. But how much? As the original manuscript is unfortunately lost, it is difficult to know whether, and to what extent, he altered it in the process of preparing the book for publication. A copy of the manuscript was listed among the sale of the estate of the Leiden naturalist Laurens Theodorus Gronovius (1730–1777), but its subsequent fate is unknown.[74] The *Ichthyologia*'s title page stated that Linnaeus had not only

69 Ibid.

70 As has been argued for Linnaeus' similarly slender work (albeit without mention of specific prices), see for example: Broberg, *Carl Linnaeus*, 272; Koerner, *Linnaeus*, 40.

71 Benjamin Schmidt, *Inventing Exoticism: Geography, Globalism, and Europe's Early Modern World* (Philadelphia: University of Pennsylvania Press, 2015), 18.

72 *Icht., Genera piscium*, sig. 2r/v and sig. *3r/v.

73 For a discussion on dedications, see: Rienk Vermij, "On the Function of Dedications in Early Modern Scientific Books," *Nuncius* 33, no. 2 (2018): 171–197.

74 An annotated copy of the auction catalogue of Lauren Theodorus Gronovius' books lists the sum for which every item was sold, but indicates that the manuscript was *'uytgehouden'*, which likely means that it was excluded from sale. *Catalogus librorum exquisitissimorum* [...] *Laurentius Theodorus Gronovius* (Leiden: Th. Haak & socios, nec non J. Meerburg, 1778) at the Koninklijke Bibliotheek (hereafter KB), The Hague, Verzcat. 4612, items 169 and 170 on page 177. See also: Sten Lindroth, "The Two Faces of Linnaeus," in *Linnaeus, The Man and His Work*, ed. Tore Frängsmyr (Berkeley: University of California Press, 1983), 176. At least two other copies of the manuscript were made, one for Clifford

saved it from oblivion, but had also examined, adapted and edited it[75] – though not entirely by himself, as he solicited editorial assistance from the medical student Tiburtius Kiellman (1718–*c.*1743).[76] From the inventory of Artedi's possessions that an Amsterdam notary drew up after his death we learn that though Linnaeus did adapt the titles both of the book as a whole and its constituent parts, these parts did, in fact, exist as discrete entities.[77] Some scholars have argued that as Linnaeus mentions in his biographical note that Artedi had read his entire manuscript to him, Linnaeus impact on the work was actually rather minimal.[78] If Linnaeus' statement were true, the manuscript must have been as good as finished. But Linnaeus' editorial hand is unmistakably visible in the book whenever he refers to his own *Systema naturae* (Leiden, 1735) and *Fundamenta botanica* (Leiden, 1736), which had not yet appeared at the time that Artedi completed his manuscript. In certain passages Linnaeus explicates his own ideas: "In this part Mr. Linnaeus has acted on his own, since he very recently laid this very axiom in his *Botanica*, which to the greater part is valid in Ichthyology, a few things having been changed, excluded, or added."[79] All in all, however, Linnaeus appears to have made additions to, rather than any real changes in, the system.

This chapter is not designed to settle the matter of who, precisely, contributed what to Artedi's *Ichthyologia*, but rather to understand how it made an effort to demarcate ichthyology as a separate field of natural history by considering the work both with regard to its content and its paratexts. As said before, the book's modest looks are deceiving. The hefty claims it makes begin with the title. The aforementioned inventory indicates that Artedi had given

---

and the other for Wishoff, but these have also been lost, see: Hans Aili and Theodore W. Pietsch, "Jacob Theodor Klein's Critique of Peter Artedi's *Ichthyologia* (1738)," *Svenska Linnésällskapets Årsskrift* 97 (2014): 62.

75 "Vindicavit, Recognovit, Coaptavit & Edidit Carolus Linnaeus", *Icht.*, title page.

76 Broberg, *Carl Linnaeus*, 136; Aili and Pietsch, *Peter Artedi: Reformer of 18th Century Zoology*, vol. 1, 95–96.

77 Notary minutes, Stadsarchief Amsterdam (hereafter SA), Amsterdam, Archief van de Notarissen, 344 Salomon Dorper, 10695, 30 September, 1735, no. 91. The names of these parts were *Historia literaria ichtiologiae* (probably corresponding to the first part of the printed book) *Prolegomena Institutionum manuscript* (the second part), *Synonymologia manuscript* (the fourth part) and the *Historia piscium universalis manuscript* (the third and/or fifth part). See: Hendrik Engel, "Some Artedi Documents in the Amsterdam Archives," *Svenska Linnésällskapets Årsskrift* 34 (1951): 56–57.

78 Lönnberg, *Peter Artedi*, 36, cf. Linnaeus, "Vita authoris," *Icht.*, sig. *5v.

79 As cited in Aili and Pietsch, "Jacob Theodor Klein's Critique," 47. Original quote *Icht.*, *Philosophia*, aphorism 189, at 64. The term "aphorism" will henceforth be abbreviated as aph., as is common in Linnaean studies. See also: Aili and Pietsch, *Peter Artedi: Reformer of 18th Century Zoology*, vol. 1, 58.

his manuscript the general title *Institutiones ichthyologiae*.[80] This decision is telling for various reasons. First, it uses the term *ichthyologia*, which was not a common practice at the time; most natural historical works on fish used the term *historia piscium*, and on those few occasions that *ichthyologia* did find its way into a book's title, the term itself was not explicitly addressed or explained.[81] Artedi defined ichthyology [*ichthyologia*] as "the science [*scientia*] that first specifically indicates all parts of the Fish, subsequently exposes the true Genera and Species names, and finally commemorates the noteworthy Characteristics observed."[82] The term *scientia*, too, was laden with meaning, as it denoted causal, certain knowledge.[83] That, decades later, the article on 'Ichthyologie' in the *Encyclopédie* of Denis Diderot (1713–1784) opened with this triad of indicating and naming parts, naming species and genera, and exposing particularities, attests to the durability of Artedi's definition.[84] A detailed breakdown of what precisely these steps entailed will be provided in the following section.

The second striking thing about the title chosen by Artedi, is that by using 'institutions' [*institutiones*], he underlined that his work was meant to serve as the foundations for this emerging field. The term *institutiones*, after all, had been used to refer to the foundations of learning of a certain branch of knowledge, such as law in Ancient Rome, and was applied to other branches of learning in the early modern period. The aforementioned de Tournefort, for example, published the *Institutiones rei herbaria* (Paris, 1700), and the Dutch physician Herman Boerhaave (1668–1738) gave his medical series the

80 The title of the published work was thus not invented by Linnaeus. Engel, "Some Artedi Documents in the Amsterdam Archives," 56. The contents of this manuscript have been reproduced in Orvar Nybelin, "Tvenne Opublicerade Artedi-Manuskript," 58–77. For a comparison of Artedi's and Linnaeus' titles, see: Aili and Pietsch, *Peter Artedi: Reformer of 18th Century Zoology*, vol. 1, 46–47.

81 See, for instance, Carolus Figulus, *Ichthyologia seu dialogus de piscibus* (Cologne: Eucharius Cervicornus, 1540) and Stephan Schoenevelde, *Ichthyologia et nomenclaturae animalium marinorum, fluviatilium, lacustrium* (Hamburg: ex Bibliopolio Heringiano, 1624).

82 "Ichthyologia est scientia, quae primum omnes Piscium partes nominatim indicat, deinde nomina Generica et Specifica vera ostendit, & denique Proprietates quasdam observatu dignas interdum commemorat." *Icht., Philosophia*, aph. 5, at 2. Original in emphasis.

83 Pomata and Siraisi, *Historia*, 10.

84 "L'affaire de l'Ichtyologie est premierement de distinguer toutes les parties des poissons, par leurs noms propres ; secondement, d'appliquer à chaque poisson ses noms génériques et spécifiques, c'est-à-dire ceux qui constituent son genre et ses espèces ; troisiemement d'exposer quelques-unes des qualités particulières de l'animal." Louis de Jaucourt, "Ichthyologie," in *Encyclopédie ou Dictionnaire raisonné des sciences, des arts et des métiers*, vol. 8, eds. Denis Diderot and Jean le Rond d'Alembert (Neufchâtel: Samuel Faulche, 1765), 483.

title *Institutiones medicae* (Leiden, 1708). Artedi owned an edition of the latter work.[85] His choice of title thus suggests that Artedi longed to give the study of fish a similar foundational and standardised footing. Even if the original manuscript remains elusive, therefore, its title already offers valuable clues as to its author's aspirations. The remainder of this chapter will depart from the printed version of the *Ichthyologia*.

Similar to Linnaeus' *Bibliotheca botanica* (Amsterdam, 1736), the first part of Artedi's monograph is entitled *Bibliotheca ichthyologica*. It offered an *historia literaria ichthyologiae*, a bibliographic overview of ichthyological learning.[86] The genre of the *historia literaria* emerged in the context of sixteenth-century humanistic learning, and entailed a history of all human knowledge recorded in writing.[87] Artedi applied the genre of bibliographic overview to the field of ichthyology, like Linnaeus did for botany.[88] Both works offered insight not only into how these naturalists envisaged the past of these fields of natural historical knowledge, but also their present status and even future directions.[89] The overview was preceded by an index in which the various authors who had written on fishes were arranged neatly by time period, from the centuries before Christ up until the moment of Artedi's own writing. In this way, Artedi put himself as the newest in an old and venerable tradition, while indicating that he ushered in a new phase in the study of fishes.

In his bibliographical overview, Artedi briefly summarised the contents of each book he listed and evaluated its merits and pitfalls, discussing its structure, the quality of the images, as well as the style. These book reviews, however concise, offer illuminating insight into his beliefs about how natural history should be done. Several things stand out. First of all, he paid particular attention to whether books displayed a certain method. Artedi distinguished between those natural histories of fishes that were based on some form of method, and those histories without. In assessing the work of others, for example, he uses 'method' for any organising principles that they may deploy: fish could

85 SA, 344 Salomon Dorper, 10695, 30 September, 1735, no. 91.

86 How Artedi managed to consult all these books remains unclear. The inventory drawn up after his death includes the works of Jan Jonston and Willughby and Ray, and were probably on loan from Seba.

87 Michael C. Carhart, "Historia Literaria and Cultural History: From Mylaeus to Eichhorn," in *Momigliano and Antiquarianism: Foundations of the Cultural Sciences*, ed. Peter Miller (Toronto: University of Toronto Press, 2007), 186–187.

88 Gessner also listed earlier authors in his natural historical works, but he did not annotate them as extensively as Artedi did; I thank Sophia Hendrikx for sharing this with me.

89 Alix Cooper, *Inventing the Indigenous: Local Knowledge and Natural History in Early Modern Europe* (Cambridge: Cambridge University Press, 2007), 156.

be grouped according to their external shape (Willughby and Ray), the kind of water in which they dwelled (Rondelet), according to the alphabet (Gessner), or no method may be used at all, and the fish themselves positioned at random (Salviani).[90] He noted, furthermore, if authors had, rather than drawing from their own observations, transcribed the descriptions of others (Jonston, Aldrovandi).[91] In discussing Gessner, he qualified the naturalist's style as discursive, "in the manner of the ancients."[92] Artedi contended that descriptions ought to be short and succinct because diffuse and long descriptions served no purpose. After all, he declared, the one and only goal of natural history was the discovery of the genera and species of *rerum creatarum*, created things.[93]

On the whole, little is known about Artedi's religious ideas, but his use of the phrase *rerum creatarum* does point to the physico-theological underpinnings of his work. He believed that the task of the naturalist was to uncover God's divine blueprint by finding patterns and order in created beings: in the case of fishes, this meant looking carefully at their physical features, such as the number and position of their fins and teeth. Artedi took a quantifying approach in which number, order and structure were key concerns – very much in the spirit of early-eighteenth century science. The system that emerged from this, Artedi believed, would then reflect the actual structure that God had unfolded, and thus constitute a true and natural system. He does not explain what precisely made a system natural and true. The search for the 'natural method' in arranging plants and animals was a discussion of much longer standing and continued to be hotly debated in the eighteenth century.[94] The natural method entailed grouping plants alongside each other according to their morphological similarities, and it was widely held that by employing it one could uncover the blueprint of Creation.

Artedi referred to his classification system as a 'natural method', which he equated with the 'true' method, one which stood in opposition to an 'artificial

90 *Icht., Bibliotheca ichthyologia*, 64–65.

91 Ibid., 63.

92 "more veterum." Ibid., 30. Incidentally, the year of publication that Artedi gives for Gessner's *Thierbuch* (1558) is incorrect.

93 "[...] diffusae & longae descriptions proprietatum & qualitatum in Ichthyologia & reliqua Historia Naturali inutiles sunt, quatenus vera & naturalis methodus in dignoscendis Generibus & Speciebus rerum creatarum, unicus & praecipuus finis Historiae Naturalis fit." *Icht., Philosophia*, aph. 6, at 2.

94 For the background to this discussion in the seventeenth and eighteenth centuries, see: Scharf, "Identification Keys," 86–91; Lesch, "Systematics and the Geometrical Spirit," 76–80; and James L. Larson, "Linnaeus and the Natural Method," *Isis* 58, no. 3 (1967): 304–320.

method', which he deemed 'false'. The 'natural' method rested on markers strictly defined by physiological features such as "number, shape, position, and proportion of body parts," whereas the artificial method rested on extrinsic features such as habitat, diet, and climate.[95] These latter features were of central importance for vernacular, practical knowledge of fish. To be able to grasp fish in this way required a sustained knowledge of fish as living, breathing creatures (be it through gills or otherwise) swimming in seas, rivers and streams alongside other fish and aquatic creatures – a type of knowledge that we might, in contemporary terms, call ecological. Artedi was interested in an altogether different type of interrelation: his aim was to find the answer to how different species of fish could be placed in a hierarchical system, the key to which was hidden in their body parts.

As he did for the many fish that he dissected, Artedi severed the existing body of knowledge about fish into its component parts and examined their uses. While he acknowledged that there were different authors who had written about fish, doing so for various reasons and to diverse ends, he declared that his work would omit writings from "Physicians, Chemists, Physicists, Mechanics, Oeconomists, and of those who wrote of Culinary matters or about Fishery."[96] He does not specify what these categories of fishy knowledge entail, but only qualifies them as being *amethodicis*, or 'without method', and therefore not meriting a place in his ichthyological work. Artedi's explicit dismissal of knowledge with regard to culinary matters or fishery constituted a clear move away from earlier traditions of learned inquiries into fish. As we have seen in the previous chapters, natural historical volumes that discussed fish published in the sixteenth and seventeenth centuries encompassed remarks on the culinary uses of mackerel or longer discourses on subjects such as the curing herring for trade and commerce, as found in *Historia piscium*. Chapter 2 has shown how fishermen distinguished species by their physical differences and how a good deal of their identifications were taken on in the *Historia piscium*.

Such examples show how theory and practice defy simple binaries. Artedi's written sources, of which the *Historia piscium* was the most important, were based on exchanges with people whose experience was rooted in practice. Stepping out of the library and on the lookout for new species to dissect, Artedi moved in a world where fish were inextricably part of culinary life and medical

95 Pietsch and Aili, "Swedish Naturalist Peter Artedi (1705–1735) and His Place in the History of Biosystematics," 591.

96 "Alii vero pro varia intentione & fine aliter quoque de Piscibus agunt, ut Medici, Chemici, Physici, Mechanici, Oeconomi, Coquinarie et rei Piscariae scriptores, ut jam de Amethodicis nihil dicam." *Icht., Philosophia*, aph. 6, at 2.

custom, hauled up in nets to be sold in market stalls in great number. In some instances, his writings show that Artedi was interested in how fishermen distinguished between species of seal, how the flesh of perch was white and tasty, and that the grayling spawned in May.[97] By and large, however, this context of time and place is absent in his book. Ultimately, Artedi intended to introduce a descriptive format that was in full service to his classificatory method, as will be discussed in detail in the next section. Rather than a complete rejection of practical and ecological sources, therefore, we should see it as an expression of Artedi's ideas of how to separate and classify knowledge about fish itself.

Like Artedi, Linnaeus attempted to organise and classify those people who inhabited his world, though of course this was the world of plants, and included gardeners and horticulturists.[98] For the botanist, he drew up the following job description: "he who knows to call similar vegetables with similar names and distinctly different plants with distinctive names, intelligible to everyone."[99] In his *Fundamenta botanica* (Leiden, 1736) Linnaeus had distinguished between *botanici*, who operated according to the systematic principles of botany, and *botanophili*, whose engagements with plants were not based on the fundaments of botany, such as anatomists, gardeners and physicians.[100] Here it is pertinent to note that Linnaeus himself was involved in horticulture and that this garden work informed his theoretical, taxonomical thinking, again indicating that in truth these apparently distinct areas of knowledge were not fenced off from one another.[101] Both Artedi and Linnaeus demarcated their particular kind of learned, systematised knowledge from practical or artisanal knowledge. They also set natural historical knowledge apart from the kind of knowledge they considered the prerogative of physicians and anatomists: in discussing the various parts of fish, for example, Artedi remarked that the study of their blood vessels and nerve systems did not properly relate to ichthyology, but rather to the domain of comparative anatomy, and that it, therefore, did not require

97 Aili and Pietsch, *Peter Artedi: Reformer of 18th Century Zoology*, vol. 1, 156–157; Artedi, *Icht., Descriptiones specierum piscium*, 43, 80–81.

98 Hodacs, Nyberg and Van Damme, "Introduction," 3.

99 Linnaeus, *Genera plantarum*, 1, as translated in Staffan Müller-Wille and Karen Reeds, "A Translation of Carl Linnaeus's Introduction to *Genera plantarum* (1737)," *Studies in History and Philosophy of Biological and Biomedical Sciences* 38, no. 3 (2007): 265.

100 "BOTANOPHILI (6) sunt, qui vegetabilia, licet non ex fundamentis Botanicis (151), tractant, ut: *Anatomici, Hortulani, Medici, Anomali.*" Carl Linnaeus, *Fundamenta botanica* (Amsterdam: Salomon Schouten, 1736), aph. 43, at 4.

101 See, for example: Stearn, "Carl Linnaeus," 21–31.

discussion in his book.[102] What all of this makes abundantly clear was that, for Linnaeus and Artedi, not anyone who busied themselves with plants or fish was to be considered a botanist or an ichthyologist, respectively.

The orderly manner in which the contents of the *Ichthyologia* are structured, not least the regular use of lists, is of particular note. The authors discussed in the *Bibliotheca ichthyologia*, for instance, are divided into methodic authors and a-methodic authors.[103] Artedi's reviews of their publications take the form of consecutively numbered lists. While lists had figured in natural historical studies for centuries, the rigorous way in which they are applied in Artedi's book is remarkable, as almost every part of it takes the form of a list, table or index: not only the descriptions of genera and species of fish, but also the people writing and otherwise engaged with fish. James Delbourgo and Staffan Müller-Wille have characterised lists as tools that "simultaneously inventoried and organised the accumulated world."[104] To this it can be added that, besides offering a practical way for overwhelmed naturalists to keep track of the abundance of knowledge available to them, the making of lists also gave them an opportunity to actively shape the corpus.

Other tried and tested ways of structuring knowledge could be found in scholastic works. In explaining his principles for the study of fish, Artedi presents them in the form of a *definitio* followed by a *scholion*: a definition of a certain concept and a commentary to that definition. This is the structure of the *disputatio* that was very common to the scholastic method that still predominated university curricula.[105] In other instances, he offered his principles in short propositions which he called *theses*, and which he then elaborated upon with a *demonstratio* or *observatio*; such reasoning structures stretch back to traditional Euclidean forms of argumentation.[106] In presenting his new approach to fish, Artedi drew on traditional ways of organising and conveying knowledge, such as the *historia literaria* and the *disputatio*, to get across his method and place.

---

102 "Descriptio Arteriarum, Venarum, Vasorum Lymphaticorum & Nervorum in Piscibus, non proprie ad Ichthyologiam seu Historiam Piscium naturalem, sed ad Anatomiam Comparatam pertinet [...]" *Icht., Philosophia*, aph. 113, at 44. Original in emphasis.

103 Linnaeus also arranged the authors discussed in his *Bibliotheca botanica*, see: John L. Heller, "Linnaeus's *Bibliotheca Botanica*," *Taxon* 19, no. 3 (1970): 363–411, esp. 365.

104 James Delbourgo and Staffan Müller-Wille, "Introduction: Listmania," *Isis* 103, no. 4 (2012): 713.

105 For a historical overview, see: Alex J. Novikoff, "Toward a Cultural History of Scholastic Disputation," *American Historical Review* 117, no. 2 (2012): 331–364.

106 Tore Frängsmyr, "The Mathematical Philosophy," in Frängsmyr, Heilbron and Rider, *The Quantifying Spirit*, 39.

In short, the *Ichthyologia* is a unique document showing how one could demarcate ichthyology as a separate field of knowledge. In it, Artedi drew up the first definition of ichthyology, and constructed, surveyed and commented on its history. He did so with a good sense of scholarly custom, placing this new subdiscipline within the best learned traditions. Central in all these aspects was the notion of 'method'; which in its broad sense entailed an orderly and systematic approach, and its narrowest sense referred to the 'natural method.' Despite the ambiguities of the term prevalent in Artedi's work, the overall idea was that any method was better than no method, and that *his* method was the best. In the process, Artedi excluded certain types of knowledge and their practitioners, whittling away their authority in the process. This ambition was present in his book's content as much as in its structure: it listed what was important. So far, the main tenets of Artedi's system have only been discussed briefly. It will now be discussed more thoroughly, both because it was consequential for the natural historical study of fish, and because it demonstrates how Artedi's principles worked in practice.

3 Classifying Fish

What ought an ichthyologist do? According to Artedi, instilling *order* into the world of fish should be the principal aim. To do so, clear concepts and categories were required. These he offered in the second part of his book, the *Philosophia ichthyologia*. Artedi laid out an elaborate set of rules to which the ichthyologist should adhere. These rules explained, for example, how one could demarcate fish from other classes of animals, how one might first group them into orders and subsequently these orders into genera, and how one unambiguously distinguished one species from the other. Artedi also proposed a reformation of genus and species names. His focus on clear and distinct categorisation of species also entailed a process of abstraction, as we will see, in which focus was on measuring and counting and the historical, qualitative component of natural history came to matter less.

The first step in classifying fish was to decide what a fish actually was. Artedi described that there were six general classes in zoology: the hairy quadrupeds, the birds, the amphibians, the fish, the insects and the zoophytes.[107] Each of these classes formed their own field of study, which were, respectively, "the natural history of hirsute animals, ornithology, amphibiology, ichthyology and

107 *Icht.*, *Philosophia*, aph. 134, at 49.

entomology."[108] Artedi defined fish as "an animal without feet, always endowed with fins, breathing either through gills or lungs, dwelling mostly in water, there swimming either only with its fins, or simultaneously flexing the body, sometimes willingly venturing onto land, and in some instances flying in the air above the water with the help of pectoral fins."[109] Comparing this definition to the one drawn up by Willughby and Ray, as cited in Chapter 1, some things stand out: both studies label a fish as an aquatic animal lacking feet, but where Willughby and Ray hold that a fish never comes out onto dry land, Artedi contends that certain fish sometimes venture ashore out of their own volition, and also that one may also encounter them flying in the air.

After having defined what ichthyology is, and what a fish is, Artedi proceeded to describe each and every part of a fish. In doing so, he wanted to offer an "explanation of the technical terms common to ichthyology" – that is, a consistent terminology and vocabulary that could henceforth be applied in the describing of fish.[110] He discussed what, among other parts, fins, tails, scales, and gills, as well as stomachs or swim bladders, looked like in different taxonomical groups of fish, and also indicated their worth for classification. Fins, he explained, were those parts of the body of fish that protruded from the body and that had membranes that could either be soft or thorny, the latter being the fin rays.[111] Although, as we saw in Chapter 1, Willughby and Ray every now and then counted the number of rays in the fins of fish, they did not use it as one of their main taxonomical criteria, whereas Artedi saw them as key characteristics for assigning fish to both orders and genera. A proper understanding of each of part of the fish helped in the classification of species. This section dives into Artedi's classification system in more detail, discussing his taxonomical ranks in turn, from orders to genera to species, and finally, varieties. It offers us insight into the various parts from which his system was composed.

---

108 "[...] ut in 1°. in Historiam Naturalem Animalium Quadrupedum Pilosorum. 2°. In Amphibiologiam seorsim. 3°. Ornithologiam. 4°. in Ichthyologiam & 5°. in Entomologiam." Ibid., aph. 132, at 48–49.

109 "Piscis est Animal apodum, Pinnis semper praeditum; vel branchiis, vel Pulmonibus respirans; plerumque in aqua habitans, ibique vel solis Pinnis, vel flexuoso corporis impulsu simul natans, interdum vero in terram sponte egrediens, & quandoque in aëre supra aquam ope pinnarum Pectoralium volans." Ibid., 1–2. Original in emphasis.

110 *Icht., Praefatio authoris*, sig. **r.

111 *Icht., Philosophia*, 3. For fin rays, Artedi introduced the literal term *ossiculum* [small bone] for the hithertho broadly used term *radius* [staff or rod], that rested on metaphor. Pietsch and Aili, "Swedish Naturalist Peter Artedi (1705–1735) and His Place in the History of Biosystematics," 594.

3.1 *Orders*

Now that the class of 'fish' had been defined, the next step was to divide it into orders. Artedi thought that the existence of this taxonomical level should be obvious to everyone. After all, where certain genera of fish were much alike, others differed greatly from one another. While the genus of the carp, for example, much resembled that of the herring, the genus of the herring and of the whale were different in nearly all of their parts.[112] From this, one could infer that the herring and the carp belonged to one and the same order. Such a subdivision into orders was useful to natural historians, because by combining species into groups of several sizes one obtained a clearer view of how they interrelated.[113] Aiming to devise a natural system, Artedi urged:

> that the *Orders* should be *Natural*; for very many and diverse Orders can be formed by Number, Form, and placement of Parts without much effort, but such orders must not be tolerated, where fish of the same kind are separated badly and diverse ones are often put together in one [Order], and indeed the square things are mixed with the round and the highest with the lowest.[114]

Those jumbled orders of dissimilar-looking animals and plants emerged when one took 'accidental things' [*res accidentales*], such as place, time of bloom or procreation, nourishment, quantity, or height as the grounds for grouping them.[115]

In order to prevent such disorder, Artedi formulated a set of strict rules that should be assumed by any naturalist. The identification key that he designed unlocked the five different orders [*clavis ordinum*] (Figure 9). It can be used as a kind of flow chart, taking the reader stepwise through several physical characteristics. At every step, the reader selects the physical characteristics of the specimen at hand, from the general to the more specific, narrowing down the options so as to eventually arrive at the correct order. Linnaeus published a similar diagram in his *Philosophia botanica*, which displays de Tournefort's key for the orders of plants in the same fashion. Both Matthew D. Eddy and Isabelle Charmantier have drawn attention to the Ramistic roots of this

112 Ibid., aph. 125, at 47.

113 Ibid., aph. 127, at 47.

114 "[…] monui ejusmodi *Ordines* debere esse *Naturales*; nam plurimi & diversi Ordines à Numero, Figura & situ Partium sine magna opera effingi possunt, sed tales ordines non tolerari debent, quatenus Pisces congeneres male separantur & diversi in unum saepe conjunguntur, adeoque quadrata rotundis & summa imis miscentur." *Icht.*, *Praefatio authoris*, sig. *v.

115 *Icht.*, *Philosophia*, aph. 128–1, at 47.

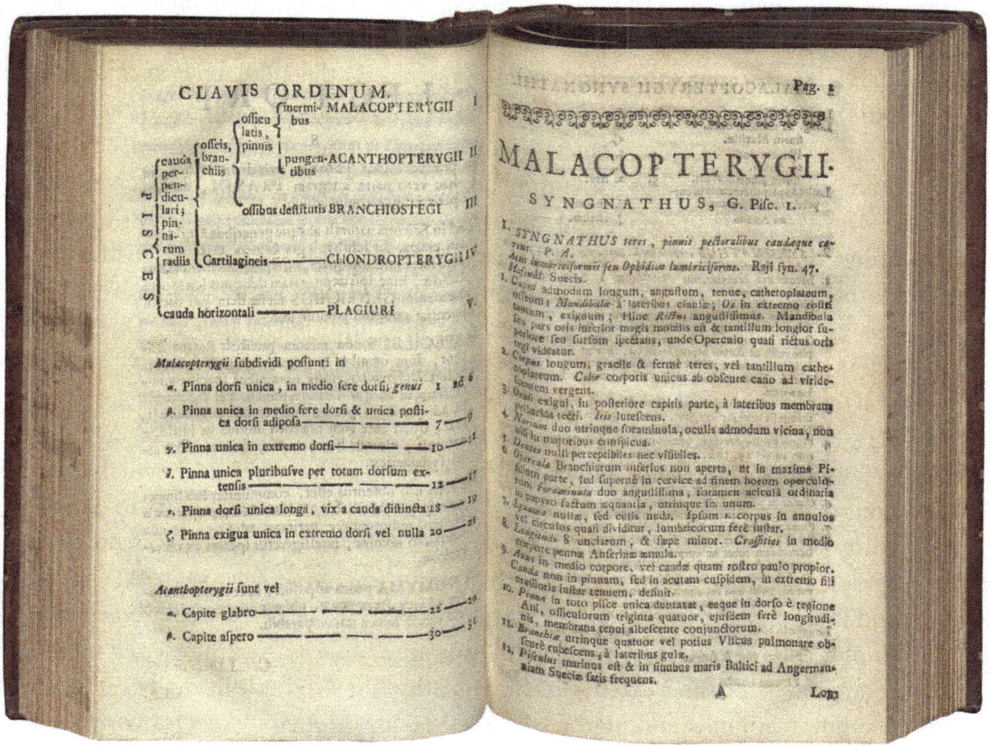

CLAVIS ORDINUM.

PISCES cauda perpendiculari; pinnarum radiis offeis, branchiis officulatis, pinnis inermibus — MALACOPTERYGII I
pungentibus — ACANTHOPTERYGII II
oſſibus deſtitutis — BRANCHIOSTEGI III
Cartilagineis — CHONDROPTERYGII IV
cauda horizontali — PLAGIURI V.

Malacopterygii ſubdividi poſſunt: in

α. Pinna dorſi unica, in medio fere dorſi; *genus* 1 ad 6
β. Pinna unica in medio fere dorſi & unica poſtica dorſi adipoſa — 7 — 9
γ. Pinna unica in extremo dorſi — 10 — 11
δ. Pinna unica pluribusve per totum dorſum extenſis — 12 — 17
ε. Pinna dorſi unica longa, vix a cauda diſtincta 18 — 19
ζ. Pinna exigua unica in extremo dorſi vel nulla 20 — 21

Acanthopterygii ſunt vel

α. Capite glabro — 22 — 29
β. Capite aſpero — 30 — 31

Pag. 1

MALACOPTERYGII.

SYNGNATHUS, G. Piſc. 1.

1. *SYNGNATHUS teres, pinnis pectoralibus caudaque carens. P. A.*
Acus lumbriciformis ſeu Ophidion lumbriciforme. Raji ſyn. 47.
Haſmal. Suecis.
1. *Caput* admodum longum, anguſtum, tenue, catheroplateum, oſſeum; *Mandibulæ* à lateribus clauſæ; *Os* in extremo roſtri ſitum, exiguum; Hinc *Rictus* anguſtiſſimus. Mandibula ſeu pars oris inferior magis mobilis eſt & tantillum longior ſuperiore ſeu ſurſum ſpectans, unde Operculo quaſi rictus oris tegi videatur.
2. *Corpus* longum, gracile & fermè teres, vel tantillum catheroplateum. *Color* corporis unicus ab obſcure cano ad virideſcentem vergens.
3. *Oculi* exigui, in poſteriore capitis parte, à lateribus membrana pellucida tecti. *Iris* luteſcens.
4. *Narium* duo utrinque foraminula, oculis admodum vicina, non niſi in majoribus conſpicua.
5. *Dentes* nulli perceptibiles nec viſibiles.
6. *Opercula* Branchiarum inferius non aperta, ut in maxima Piſcium parte, ſed supernè in cervice ad finem horum operculorum, *Foraminula* duo anguſtiſſima, foramen aciculà ordinaria in papyro factum æquantia, utrinque ſc. unum.
7. *Squamæ* nullæ, ſed cutis nuda. Ipſum v. corpus in annulos vel circulos quaſi dividitur, lumbricorum ferè inſtar.
8. *Longitudo* 8 unciarum, & ſæpe minor. *Craſſities* in medio corpore pennæ Anſerinæ æmula.
9. *Anus* in medio corpore, vel caudæ quam roſtro paulo propior. *Cauda* non in pinnam, ſed in acutam cuſpidem, in extremo fili craſſioris inſtar tenuem, deſinit.
10. *Pinna* in toto piſce unica duntaxat, eaque in dorſo è regione Ani, oſſiculorum triginta quatuor, ejuſdem ferè longitudinis, membrana tenui albeſcente conjunctorum.
11. *Branchiæ* utrinque quatuor vel potius Viſcus pulmonare obſcurè rubeſcens, à lateribus gulæ.
12. *Piſculus* marinus eſt & in ſinubus maris Baltici ad Angermaniam Sueciæ ſatis frequens.

A Long

FIGURE 9 *Clavis ordinum.* Peter Artedi, *Ichthyologia, sive opera omnia de piscibus* (Leiden: Conrad Wishoff, 1738), n.p.
LEIDEN UNIVERSITY LIBRARY

particular way of organising knowledge.[116] The French humanist Peter Ramus (1515–1572) had proposed a logical method of structuring knowledge according to dichotomous divisions, set out in tables or diagrams, and his teachings remained part of the curriculum of some European universities in the time of Artedi and Linnaeus.[117] We can encounter the organisational method in the works of both.

To decide to which order a certain fish belonged, the first thing one needed to look for was the position of its tail. Firstly, any fish with a horizontal tail belonged to the order of cetaceans (*Plagiuri*). Those fish with perpendicular tails

116 Isabelle Charmantier, "Carl Linnaeus and the Visual Representation of Nature," *Historical Studies in the Natural Sciences* 41, no. 4 (2011): 370–371; Matthew D. Eddy, "Tools for Reordering: Commonplacing and the Space of Words in Linnaeus's *Philosophia Botanica*," *Intellectual History Review* 20, no. 2 (2010): 227–252, esp. 243.

117 On Ramus and his influence, see: Walter Ong, *Ramus, Method and the Decay of the Dialogue: From the Art of Discourse to the Art of Reason* (Cambridge, Mass.: Harvard University Press, 1958).

could be divided into those with cartilaginous skeletons (*Chondodropterygii*) and those with bony skeletons. That latter group, in turn, could be further subdivided in those fish that did not have any rays in the gill flap (*Branchiostegi*) and those that did possess such rays.[118] The last dividing feature was whether a fish was 'unarmed' (*Malacopterygii*) or 'eager to fight' (*Acanthopterygii*); rather than implying something about the temper of the fish, this indicated whether the rays in its fins were soft or thorny.[119] The names that Artedi gave his five orders of fishes are, in fact, Greek descriptions naming the primary distinctive feature of each order (these were, in the same sequence as just followed: flat fins, cartilaginous fins, large gill lids, soft fins, and thorny fins).[120] One word thus sufficed to place each fish in its proper order.

3.2 *Genera*

Below the order, one entered the rank of the genus. Chapter 1 has shown that the notion of 'genus' or its plural, 'genera' had been used since Antiquity to denote groups of plants or animals that resembled one another. There was no clear agreement, however, on *what*, exactly, a genus was. As we have seen, Willughby and Ray applied the term rather loosely to indicate species that shared certain similar external characters. While they grouped species together based on the resemblance of their external characteristics, they did not formally establish taxonomical ranks and neither did they endow groups of species with their own names. Fish were sorted into broad or long fish. Subsequently, fish from the latter category were further described as elongated or shorter long fish.[121] Not of all Willughby and Ray's groupings were devised in such a way that they were necessarily mutually exclusive of one another: one species could belong to more than one group.[122] Thus, although the concept of genus was not unknown, it had never been clearly defined.

That Artedi considered the establishment of genera as the first and foremost aim of natural history is clear from how he, in his preface, referred to the third part of his book entirely dedicated to describing all genera: *ipsum opus*, or the work itself.[123] In order for genera to be meaningful in a hierarchical classification system, a consistently defined and applied concept of genus

118 The gill flap, also known as the operculum, is the part of the fish that covers the slits of the gills.

119 Artedi subdivides these two orders on the level of the *manipules*, a category resembling what we might now call 'families' and which he does not fully define or develop. See also: *Icht., Philosophia*, aph. 136 at 50 and Lönnberg, *Peter Artedi*, 29.

120 For this observation, I thank Hans Aili.

121 *Hist. pisc.*, 46.

122 Lönnberg, *Peter Artedi*, 27.

123 Aili and Pietsch, *Peter Artedi: Reformer of 18th Century Zoology*, vol. 1, 62.

was required. Artedi defined a genus as a group "assembled of some certain species, or resemblance of certain fish from diverse species, that *always* convene in the placing of its external parts, *usually* in number, and *often* in figure and proportion."[124] What one had to look for, therefore, was a "certain likeness between certain species, which agrees in shape, location, number or proportion of parts in such a manner that they differ from all species of the other genera in one minimal part."[125] Artedi was well aware that assigning species to their correct genera was an intricate process, and explained that general characteristics [*characteres generici*], i.e., marks that were indicators of a certain genus, had to be chosen judiciously.

He contended that even though it was important to take a good look at a species' external parts, one should at the same time not be deceived by its general outward appearance. At first glance, for example, one might be led to believe that the tench and the lake-trout belong to one and the same genus. The overall shape of these fishes was, after all, rather similar. On closer inspection, however, Artedi argued that it was plain that the fish displayed different general characteristics.[126] The three bones in the gill flap of the tench meant it had to be assigned to the genus of the carp; the lake trout had no fewer than ten such bones, and thus belonged to the genus of the salmon.[127] So far, so good. What complicated matters, however, was that general characteristics were not the same for every order of fish. This makes sense: there were no perfectly constant characters that could be used to determine each and every genus, simply because no one part was found in every fish. For the bony fishes, as we just saw, the best course of action was to count the number of bones in the gill flap.[128] For the cartilaginous fishes, the thing to pay attention to was the shape of their body and their gill apertures.[129] In the cetaceans, as the introduction to this chapter already mentioned, the salient generic characteristics were the number of fins and teeth. As every system seeking to refine and simplify matters had, inevitably, to contend with exceptions and anomalies, the genera had to be demarcated on a case-by-case basis.

---

124 "Genus Ichthyologiae est convenientia quaedam certarum specierum, seu similitudo quorundam Piscium ad speciem diversorum, qui in situ Partium externarum *semper*, numero *plerumque*, Figura & Proportione *saepe* conveniunt." *Icht.*, *Philosophia*, aph. 139, at 51. Original sentence in emphasis, which is not reproduced here for readability; the emphasis of the original, however, is retained.

125 Aili and Pietsch, *Peter Artedi: Reformer of 18th Century Zoology*, vol. 1, 4.

126 Ibid., aph. 143, at 51.

127 *Icht.*, *Philosophia*, aph. 148, at 52; see also Lönnberg, *Peter Artedi*, 31.

128 Ibid., aph. 182, at 62.

129 Ibid., aph. 153, at 54.

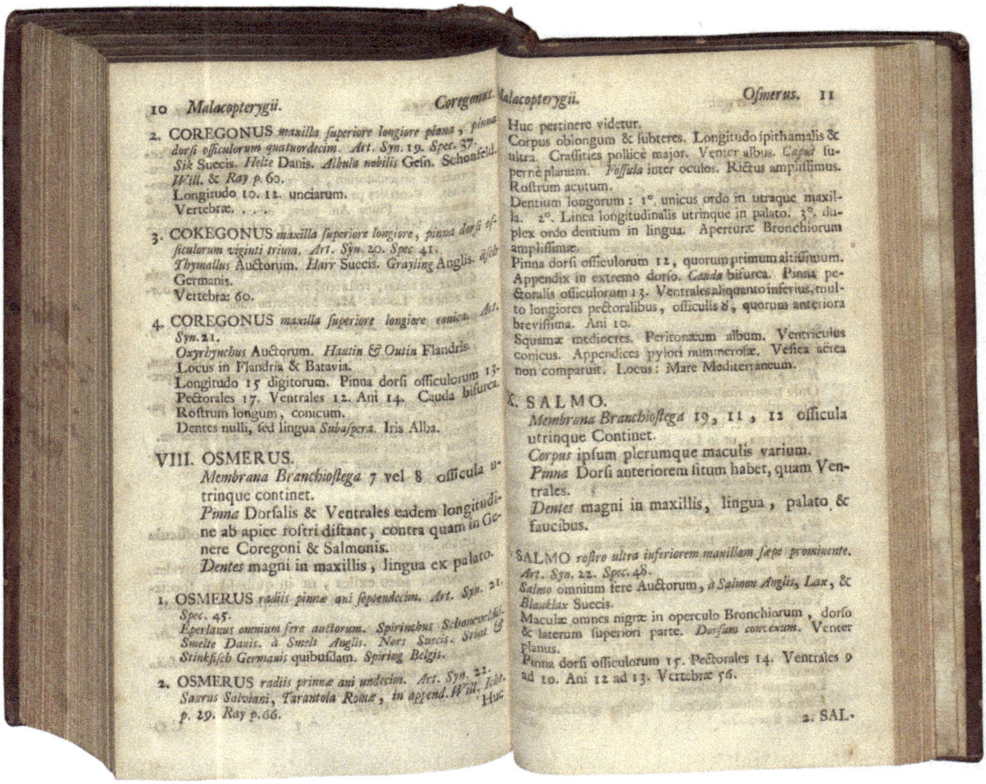

FIGURE 10 Genus description of *Osmerus*. Peter Artedi, *Ichthyologia, sive opera omnia de piscibus* (Leiden: Conrad Wishoff, 1738), 10
LEIDEN UNIVERSITY LIBRARY

After painstakingly examining this complex web of physical features, Artedi eventually arrived a grand total of fifty-two genera. These he discussed in the third part of the *Ichthyologia*, entitled *Genera piscium*. Like the rest of his book, this part is tightly structured. Organised by order, the *Genera piscium* described each genus and listed the species it comprises. A discussion of the genus called '*Osmerus*' – a type of smelt – can stand as example of his theory put to practice (Figure 10).[130] This description was given in the first order, the *Malacopterygii*. Of this order, the *Osmerus* was the eighth genus and was assigned a corresponding roman numeral (VIII). It then listed the defining physical characteristics of this genus: the bones in its gill flap numbered seven or eight, the dorsal and the ventral fins were positioned equally far from the tip of the snout, the teeth in this mouth were big, and the tongue came from the palate. Any species

130 *Icht.*, *Genera piscium*, 10.

conforming to each and every one of these characteristics could be safely assumed to belong to the genus of the *Osmerus*. This description is exemplary for how carefully selected combinations of characteristics were used to discern one genus from the other.

Confusion over genera could be prevented with a rigorous naming protocol. Artedi therefore proposed a reformation [*reformatio*] of naming practices.[131] The first rule stated that each fish belonging to the same genus must be given the same generic name. While this might sound obvious to us, Artedi showed the necessity for this measure by explaining that the various species belonging the genus *Cyprinus* (carp) were referred to by no fewer than twenty-five different generic names.[132] Artedi furthermore dictated that a generic name should consist of one word only, banning composites or diminutives.[133] Instead of vernacular (or, in Artedi's words, 'barbaric' [*barbara*]) names, only those originating from Latin or Greek were to be allowed; the ancients had offered plenty such names to draw from.[134] Yet not each Greek or Latin name met his requirements. Any similitude between the names of animals living on land and those living in the water should be avoided. Fishes were not to share their species names with mammals like the wolf [*lupus*], birds like the eagle [*aquila*], insects like the scarab [*scarabaeus*], nor vegetables like parsnip [*pastinaca*].[135] Unambiguous, clearly defined and named genera were integral to Artedi's system.

3.3 *Species*

With Artedi's key in hand, a naturalist may have decided that the fish specimen before them belonged to the order of the *Malacopterygii* on account of it having a bony skeleton and soft fin rays. Judging from the number of bones in the gill flap, the position of its dorsal and pelvic fin, and its teeth and tongue, it was part of the Osmerus genus. The question then arose precisely with which of the two species in this genus he was dealing. Artedi gave the following definition of a species:

> In *Ichthyology*, every Fish constitutes a species if it is different from the other species of its genus with regard to a certain external Part being

131 "Reformatio Nominum Genericorum" *Icht., Philosophia*, 64; "Reformatio Nominum Specificorum" ibid., 80.

132 *Icht., Philosophia*, aph. 191, at 64–65.

133 Ibid., aph. 196, at 71 and aph. 202, at 78.

134 *Icht., Praefatio authoris*, sig. **v. Linnaeus contended the same for plants: Carl Linnaeus, *Critica botanica* (Leiden: Conrad Wishoff, 1737), aph. 229, at 48.

135 *Icht., Philosophia*, aph. 194–1/6, at 66–70.

> absent or present, either as regards number, Proportion, or Shape, or owing to some constant difference in colour.[136]

This definition is an example of Kärin Nickelsen's contention that species are "not concrete objects the properties of which can be gathered just by looking at them", but rather "abstract concepts designating whole classes of objects whose definitions depend on the taxonomic system used at the time."[137] Species, then, are not self-evident entities, but they require a certain effort to be conceptualised as such. Artedi's definition can serve as a reminder that the way in which one species is demarcated from the other is a result of decisions made by the taxonomist. In this case, the taxonomical properties that Artedi asserted as relevant were number, proportion, figure, and colour.

In his letter to the reader, Artedi asserted that he had seen each and every fish described in his book for himself (with the exception of some of the whales, including, as we have seen, the siren).[138] This attests to the weight that continued to be attached to direct observation. Like Linnaeus, Artedi had built up his system through rigorous investigation of hundreds and hundreds of specimens.[139] Unfortunately for historians wishing to reconstruct his working practices, however, his species descriptions seldom mention exactly how and where he had come across the specimen at hand. An exception are the few instances where he has added 'I have seen' [*vidi*] or 'I have discovered' [*inveni*] to his descriptions.[140] It is perhaps by not detailing the particular context of his observations, that he emphasised their general nature. It implied that his observations were universal: what he saw was, in effect, the platonic form of the fish rather than a particular fish. If this indeed was Artedi's intention, it presents a compelling contrast: while Willughby and Ray asserted the reliability of their observations by specifying where and when they had seen a certain specimen, Artedi made his reliable by leaving said information out.

---

136 "Species in *Ichthyologia* appellatur unusquisque Piscis, qui a reliquis sui generis speciebus in Parte quadam externa, secundum defectum vel excessum, numerum, Proportionem, Figuram & colorem constantem variantem, diversus est." Ibid., aph. 209, at 74. Emphasis from original has been reversed.

137 Kärin Nickelsen, *Draughtsmen, Botanists and Nature: The Construction of Eighteenth-Century Botanical Illustrations* (Dordrecht: Springer, 2006), 73.

138 *Icht., Praefatio authoris*, sig. ***r.

139 Cf. Müller-Wille, "Collection and Collation," and Mary P. Winsor, "Linnaeus's Biology Was Not Essentialist," *Annals of the Missouri Botanical Garden* 93, no. 1 (2006): 2–7.

140 He mostly used 'vidi' to specify that he saw a specimen, and 'inveni' for describing the internal organs of the fish. e.g., *Icht., Descriptiones specierum piscium*, 19, 28, 35, 39, 59, 71, 79, 89, 110.

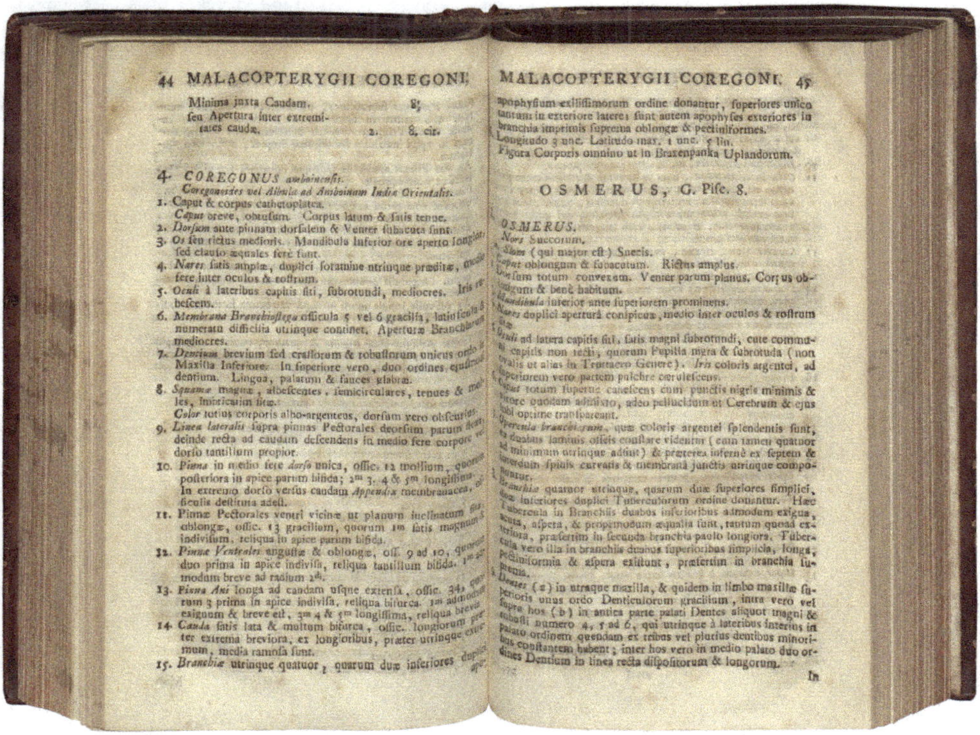

FIGURE 11 Species description of *Nors*. Peter Artedi, *Ichthyologia, sive opera omnia de piscibus* (Leiden: Conrad Wishoff, 1738), 45
LEIDEN UNIVERSITY LIBRARY

While this often makes it difficult to establish where Artedi made his observations, a letter by Linnaeus offers a rare insight. It discusses that Artedi saw a species of smelt, known in Sweden under the name of *Slom*, every spring, and that he was intent on finding out whether it constituted a distinct species from the similar-looking *Nors*.[141] The species description of the latter can give us a sense of the practical application of Artedi's system (Figure 11).[142] Every species was discussed under the header of its corresponding genus. The number of its genus (which in the case of *Osmerus* was 8) was supplied for easy reference, so one could look up the characteristics of this genus in the *Genera piscium*. The species itself was also assigned a number: the *Nors* was the first – and only species – of the eighth genus. The description opens with the capitalised genus name ('OSMERUS'), followed by a short diagnostic sentence – this is the species name, as will be discussed in more detail.

141 Linnaeus to Johan Ernst Gunnerus, 4 March 1769, UUB, L4202.

142 *Icht., Descriptiones specierum piscium*, 45.

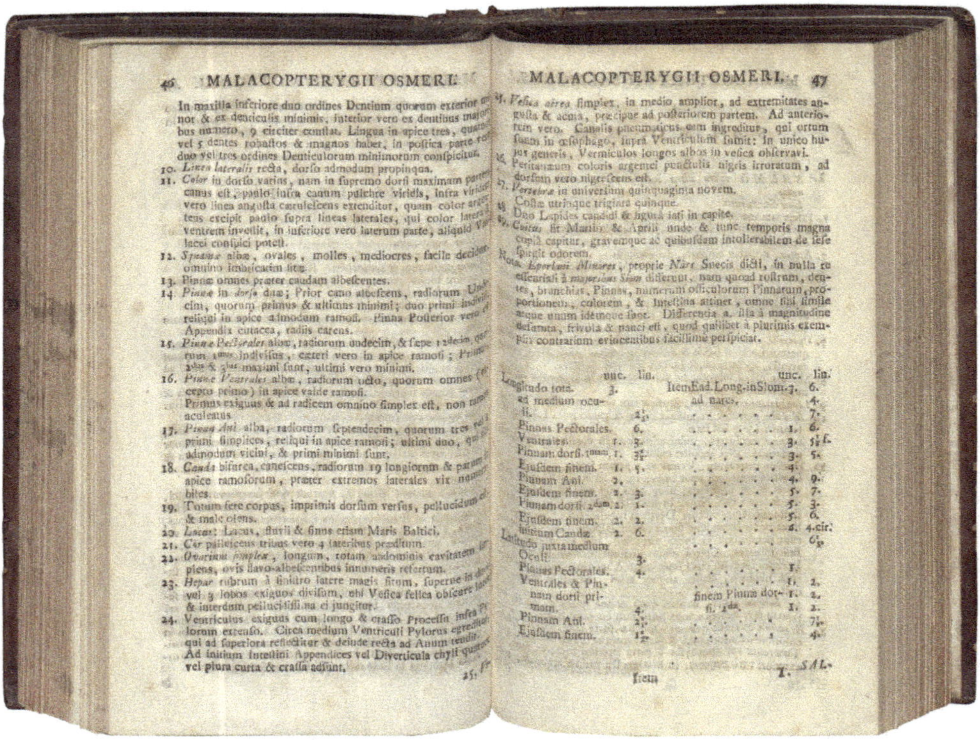
46 MALACOPTERYGII OSMERI.

MALACOPTERYGII OSMERI. 47

FIGURE 12 Species description of *Nors*. Peter Artedi, *Ichthyologia, sive opera omnia de piscibus* (Leiden: Conrad Wishoff, 1738), 47
LEIDEN UNIVERSITY LIBRARY

Artedi divided the species descriptions in numbers, "so that all [parts] are distinguished more readily."[143] The description of the smelt consisted of 29 consecutively numbered items, which documented its main attributes of the fish: for example, its gills, the gill cover, as well as each of its fins and the exact number of rays in them. He also measured the sizes of the fish's various parts with almost geometric precision in inches (unc.) and twelfths of inches (lin.). These measurements, both longitudinal and latitudinal, were formatted into a table (Figure 12). It shows that the *Nors* measured up to 3 inches long, and that the distance from the tip of its snout to the middle of its eye was 5 twenty-fourths of an inch. These measurements were given for several intervals (for example, the first dorsal fin and the second one), with the base of the tail as the last measuring point. Artedi's examination of species clearly demanded some laborious effort. Linnaeus described having seen Artedi

143 "[…] easdemque in Numeros divisi, ut omnia clariora & magis distincta essent." *Icht.*, *Praefatio authoris*, sig. **v.

"spending in many cases several whole days over one single fish" over which time he "would count over the fins, and the individual rays in them, not once only but many times […]."[144] In one breath, Linnaeus draws attention to both Artedi's careful and disciplined study of fish and his impulse to identify and classify them through quantification.

As we saw in the previous chapters, Willughby and Ray had been somewhat troubled by the abundance of names by which species could be known. They had tried to restore order to the nomenclature of fish, but did not propose a reformation of naming practices. Rather than simply restore order to the nomenclature under which fish laboured, Artedi proposed a reformation of naming practice. He contended that a species name should take the form of an "epithet, consisting of some few words, which is appended to the name of the genus, in order to distinguish one species from the other in the same genus."[145] This meant that a fish's name was comprised of a short, diagnostic summary of its defining characteristics. Take, for instance, the name Artedi gave to the common perch: *Perca lineis utrinque sex nigris pinnis ventralibus rubris*, a perch with six black fins on both sides and a red belly.[146] Artedi's naming system was somewhat more cumbersome than the binomial system that Linnaeus became well known for with the publication of the tenth edition of his *Systema naturae* in 1758. This system meant that one referred to a certain species by combining its genus name with it species name: for example, the above described perch became *Perca fluviatilis*.[147] It was decidedly less cumbersome than the nomenclature that Artedi used.[148] Nonetheless, the naming system proposed by Artedi was practical: as brief summaries of its main external features, they helped the reader distinguish between two species at a glance.[149]

However useful this new naming system was, naturalists still had to contend with the conundrum of names under which earlier authors had described species of fish. Artedi had come up with a solution for that, too. In the fourth part of his book, the *Synonymia nominum piscium*, Artedi compiled all the synonyms of fish names in various languages by means of indices. These included Latin, Swedish, Danish, German, Dutch, English, French, Italian, Spanish and Greek,

144 *Icht., Descriptiones specierum piscium*, introduction by Linnaus; translation from Lönnberg, *Peter Artedi*, 41.

145 *Icht., Philosophia*, aph. 223 at 80; translation from Lönnberg, *Peter Artedi*, 35.

146 *Icht., Descriptiones specierum piscium*, 74.

147 Linnaeus, *Systema naturae*, ed. 10, 289.

148 Linnaeus considered the binomial name to be a 'trivial name', a useful shorthand for the actual species name, the *nomen specificum*, which, as in Artedi's naming method, consisted of the generic name followed by a short descriptive phrase giving the plant's distinctive characters which set it apart from other species, see: Nickelsen, *Draughtsmen*, 80.

149 Lönnberg, *Peter Artedi*, 35.

as well as some 'miscellaneous' languages, such as Ambonese and Hispanic. Collecting these synonyms, Artedi stated, "created much work for me in comparison with the other parts of this work; for in reading so many authors about nearly every single species most of my time and my indefatigable mind were needed [...]."[150] It allowed him to showcase his comprehensive grasp of natural historical literature and his tireless industry. This part of the book functioned as a sort of fish dictionary. He also indicated where each species had been described by earlier authors, including page numbers.[151] Any naturalist wondering, for example, how the species of smooth hound shark described in the *Historia piscium* (Figure 5b) was described by Artedi could readily find out that the Swede called it a "shark with blunt or granular teeth."[152] In this way, Artedi synchronised earlier names and descriptions of fish species with those that he himself drew up for them.

3.4 *Varieties*

The last and final level of the taxonomical hierarchy were the varieties. The aforementioned species description of the smelt ('*Nors*') ends with a note addressing whether this particular species differed from the similarly looking fish (Artedi's beloved '*Slom*' that he saw each spring in Sweden):

> *The smaller Eperlani*, properly called *Närs* [sic] by the Swedes, do not differ in any essential way from *the bigger Slom*, for as far as it concerns their mouth, teeth, gills, Fins, their number of rays in the Fins, their proportion, colour, and Intestines, all are similar to each other and one and the same. The difference between them, which is selected on the basis of size, is frivolous and useless, because anyone can very easily perceive the contrary on the basis of very many evincing examples.[153]

Artedi concluded that the *Nors* and the *Slom* were the same species. For good measure, he included his calculation of the dimensions of the latter species in a table alongside that of the former (Figure 12). This example shows how ambiguities could also remain at species level. As we already saw in Chapter 1,

150 *Icht.*, *Praefatio authoris*, sig. **r/v.

151 *Icht.*, *Synonymia nominum piscium*, 1–118.

152 Ibid., 93.

153 "*Eperlani Minores*, proprie *Närs* Suecis dicti, in nulla re essentiali à *majoribus Slom* differunt, nam quod rostrum, dentes, branchias, Pinnas, numerum ossiculorum Pinnarum, proportionem, colorem, & Intestina attinet, omne sibi simile atque unum idemque sunt. Differentia a. illa à magnitudine desumta, frivola & nauci est, quod quilibet à plurimis exemplis contrarium evincentibus facillimè perspiciat." *Icht.*, *Descriptiones specierum piscium*, 47.

Willughby and Ray wondered how to decide if a certain specimen constituted a separate species, or was in fact a variation within a species. This problem was also well known to Artedi, and he lectured the reader on the matter in the second part of his book, the *Philosophia*. He stated that certain things were of little consequence for demarcating one species from the other. Fishermen, Artedi contended, distinguished between species based on matters such as the colour of the fish, whether they resided in fresh or salt water, or the time of year in which they reproduced. He deemed these kind of distinctions 'false and frivolous' [*falsa et frivolae*].[154] The colour, he expounded, of one individual fish could vary depending on its age, the season, and even the type of water it inhabited.[155] It also differed from specimen to specimen. Species could and should not be assigned on the basis of these unstable characteristics, but only on those features that he had determined to be relevant.

Because aspects such as these were too ambiguous to have meaning in any classificatory system, Artedi largely disregarded them in his species descriptions. He only occasionally mentioned the means of nutrition or procreation of a species, or the taste of its flesh, and never comments on matters relating to trade or consumption. This is what set his approach apart from earlier naturalists who had written about fish: he deliberately excluded any associations surrounding the fish, and rather focussed on the animal *in itself*, its physical presence. Zooming in on matters such as habitat or nutrition also would prove an impractical way of examining plants or animals for the naturalist who might peruse many of his or her specimens in various cabinets: these matters, after all, could not be glimpsed from looking at a specimen that had been taken out of its environment and preserved in a collection. Rather, one would have to have studied fishes for a prolonged amount of time, alive and in their own habitat. That was far from common practice for the average naturalist, however, who might combine field trips with visits to cabinets and collections. It was, in contrast, also the kind of knowledge that artisans such as fishermen, fishmongers, and cooks would be well aware of – and that, as we have seen in the previous chapter, naturalists before Artedi had made glad use of. While it is plain from the previous chapter that a fisherman's observation could be useful when it came to deciding which species was which, such individuals rarely figure in Artedi's descriptions, even though he drew on their experiences, both directly and indirectly. When using the descriptions of Willughby and Ray, for example, he was benefitting from the knowledge that these practical men had given them whether he liked it or not.

154 *Icht.*, *Philosophia*, aph. 220, at 79.

155 Ibid., aph. 213, at 75–76.

The salient characteristics were, as we saw, the number, shape and position of fins, the number of rays in the fin, or the number or shape of the teeth, or the number and shape of other parts. Artedi shared this averseness to colour with Linnaeus, for whom only 'Number, Shape, Position, and Proportion' counted in describing plants, and who railed against the – in his eyes – excessive attention of his fellow botanists to the colourisation of plants.[156] Here, too, it is worthwhile considering not only what a species description may contain, but also what has been left out. As has been discussed in the first chapter, natural historical works had long been an *encyclopedic* exercise bringing together past works and writings. Artedi's descriptions were, by and large, stripped of such historical components. While, in the *Synonymia*, he did refer to species descriptions published in earlier works, it was only with the purpose of synchronising these descriptions and identifications with his own. In so doing, he stabilised his own species identifications, subsuming and even overwriting earlier ones. His descriptions were quantitative rather than qualitative, and numerical rather than narrative.

As we have seen, his focus on order and classification had repercussions for Artedi's species descriptions, which became, in essence, simple lists of pertinent characteristics. They are examples of what Daston has characterised as the 'description by omission' in which the description of nature changed "from long accounts bristling with particulars to concise reports made deliberately bland by summary, repetition, and omission of details."[157] She locates this development between the late seventeenth and the early eighteenth century – which is also the period that separated the *Historia piscium* and the *Ichthyologia* – and a comparison of the descriptions that these works contain substantiates this hypothesis. What Artedi introduced for fish (but also for other realms of nature, as we saw) were strictly defined taxonomical ranks. In his book, he remarks that such a system of classes, orders and genera was useful not only to ichthyology, but to natural history as a whole.[158] This system would prove highly useful to later naturalists as they attempted to understand the taxonomic relations between new and already known species.

---

156 Lorraine Daston, "Description by Omission: Nature Enlightened and Obscured," in *Regimes of Description: In the Archive of the Eighteenth Century*, eds. John Bender and Michael Marrinan (Stanford: Stanford University Press, 2005), 12. Linnaeus indeed complemented Artedi's aphorisms with references to equivalent rules in the *Fundamenta botanica*.

157 Daston, "Description by Omission," 13.

158 "[…] primum et praecipuum Fundamentum, non solum Ichthyologiae, sed totius reliquae Historiae Naturalis […]" *Icht., Philosophia*, aph. 143, at 51–52.

4 Lost in Preservation?

As mentioned in the introduction to this chapter, Linnaeus presented his own classification system consisting of orders, genera and species into the first edition of his *Systema naturae*.[159] As he explained, "[i]n Ichthyology, I have not prepared my own Method, but the greatest Ichthyologist, the famous Swede mr. Petrus Artedi [...] has given me his."[160] He indeed copied the taxonomical ranks that Artedi had established with only slight alterations.[161] In the preface to the (now famous) table of orders and genera in the realm of animals of the first edition of his work, Linnaeus introduced Artedi to his readers as "the best ichthyologist of our time",[162] who had perfected the classification of fishes. While the readers could get the gist of Artedi's classification system in the present work, Linnaeus announced they could expect a more elaborate account of the underlying principles in a forthcoming publication, '*Institutiones* nempe *totius Ichthyologiae*' [*Institutions*, that is, *all of Ichthyology*] – which was, of course, eventually published as the *Ichthyologia*.[163] Where this chapter has so far analysed Artedi's *Ichthyologia* and its contents, this final section examines how Artedi's system found its way to other naturalists. It examines the consequences it had for the way in which fish were studied, and pays particular attention to discussions of whether fishes were best preserved as textual descriptions, images or objects.

We can get a better sense of how Artedi's *Ichthyologia* was received by reading the reviews that appeared in various learned journals not long after its publication. These periodicals were specifically geared towards reporting on new research and publications, and offer insight into how books were regarded by scholars.[164] French-speaking *savants* were served by the review that appeared in the *Bibliothèque germanique*, a periodical devoted to the learning of

159 Carl Linnaeus, *Systema naturae*, ed. 1 (Leiden: Theodoor Haak, 1735).

160 As translated by Broberg, *Carl Linnaeus*, 142.

161 In the order of the *Malacopterygii*, Linnaeus left out the genera of *Argentina*, *Exocoetus*, *Stromateus*, *Ophidion* and *Anableps*; in the *Acanthopterygii* order, he moved the *Blennius* genus to the *Malacopterygii* (spelling it *Blennus*), and cut the genera *Sciaena*, *Scorpoena* and *Chaetodon*. From the *Plagiuri*, he omitted the *Physeter* genus and that of the *Siren*. See: Linnaeus, *Systema naturae*, ed. 1, n.p.

162 "In *Ichthyologia* nullam ipse elaboravi Methodum, verum Suam nobiscum communicavit summus nostri temporus Ichthyologus Cl. D. *Petr. Artedi* [...]" Linnaeus, *Systema naturae*, ed. 1, n.p.

163 With the addition of 'all' [*totius*], this title was perhaps even more ambitious than the one Artedi had originally envisaged. Ibid.

164 Thomas Munck, "Eighteenth-Century Review Journals and the Internationalization of the European Book Market," *The International History Review* 32, no. 3 (2010): 417.

Germany and the countries of northern Europe,[165] while the one published in the *Nova Acta Eruditorum* catered to a Latinate audience.[166] The German polymath Johann Peter Kohl (1698–1778) wrote a glowing review of *Ichthyologia* for the *Hamburgische Berichte von den neuesten Gelehrten Sachen* in 1738, stating that the book was "arranged in a careful and provable way [...] thereby, the foundations of this science are presented with a lasting power of proof [...]."[167] He noted that "[f]or every fish, all the rays in the fin and all the vertebrae are counted; every single thing is noticed and investigated."[168] He concluded that "[...] the work has so cleared up this part of Natural History, otherwise the most difficult of all, to such a degree that one must be astounded."[169] Artedi would probably have been pleased with this critique; the words the reviewer uses to characterise his system, such as 'provable' and 'foundational', mirror the terms that he himself applied to it.

Other indications of how Artedi's work was received and used, and by whom, can be found in the correspondence between both well and lesser-known eighteenth-century naturalists. Paul Heinrich Gerhard Möhring (1710–1792) for example, inquired with Linnaeus from Jever when the work would be available and how much it would cost.[170] Christian Gottlieb Ludwig (1709–1773) asked him to send him a copy to Leipzig, so that he could pass it on to Jacob Theodor Klein (1685–1759).[171] Alexander Garden (1730–1791) affirmed to have used both Linnaeus' *Systema naturae* and Artedi's *Ichthyologia* while describing species of fish that were new to him in and around Charlestown in the province of Caroline, one of Britain's colonial territories in North America.[172] In a letter to Arnout Vosmaer (1720–1799) in October 1751, Laurens Theodorus Gronovius promised to show him his arrangement of the genera of Dutch fishes, organised

165 Jacob Theodoor Klein, "Peter Artedi, *Ichthyologia*," in *Bibliothèque germanique ou Histoire littéraire de l'Allemagne et des pays du Nord*, vol. 45, ed. Paul-Emile de Mauclerc (Amsterdam: Pierre Humbert et fils, 1739), 143–159. The attribution of this review to Klein, a translation of the Latin review that he proffered Sloane, has been done by Pietsch and Aili, "Jacob Theodor Klein's Critique," 53.

166 "Petri Artedi," in *Nova acta eruditorum*, ed. Friedrich Otto Mencke (Leipzig: Johann Friedrich Gleditsch, 1741), 652–657.

167 Johann Peter Kohl, "Opera posthuma ichtyologica," *Hamburgische Berichte von den neuesten Gelehrten Sachen* 16 (1738): 131–132. As translated in Pietsch and Aili, "Jacob Theodor Klein's Critique," 57.

168 Pietsch and Aili, "Jacob Theodor Klein's Critique," 57.

169 Ibid.

170 Paul Heinrich Gerhard Möhring to Linnaeus, 4 October 1737, UUB, L0217.

171 Christian Gottlieb Ludwig to Linnaeus, 22 October 1737, UUB, L0210. For Klein's assessment of the work, see Aili and Pietsch, "Jacob Theodor Klein's Critique," 39–84.

172 Alexander Garden to Linnaeus, 12 April 1762, UUB, L2902.

according to Artedi.[173] From Montpellier, Antoine Gouan (1733–1821) wrote to Linnaeus that he found Artedi's system useful, although he proposed some small changes.[174] When Klein received the work from Ludwig, he wrote a review of it that he then sent to Sloane.[175] He noted with some surprise that the seahorse had been assigned to the order of *Malacopterygii*, even though it lacked the perpendicular caudal fin this order required.[176] He did agree, however, with the premise and structure of the system itself.

For a considerable number of naturalists, Artedi's classification system became a common point of reference. The lists of genera and species that both Linnaeus and Artedi compiled were used to catalogue collections of *naturalia*, as the following chapter will discuss in more depth. Eighteenth-century naturalists, ever more inundated with specimens hitherto undescribed, found a welcome source of order in these systems. Linnaeus' binomial names, in particular, made it easier to compile data on plant and animal species, to produce clear lists of specimens. This is probably why the British naturalist and explorer Joseph Banks (1743–1820) had 'updated' his copy of the *Ichthyologia*, which formed part of the library he took on board with him on his voyage to Iceland in 1772, by adding the binomial names of Linnaeus.[177] Linnaeus did the same with his own copy of the *Ichthyologia*.[178] Just as Artedi had imposed his new standard name onto the previous ones existing for a species, therefore, Linnaeus and his adherents synchronised his carefully constructed species names with the Linnaean binomial names.

If all that was needed was to summarise nature's multifarious productions in a kind of linguistical matrix, was there any need to still bother with making illustrations? As has been mentioned, Artedi's book encompassed none. He does not explicitly address this lack of illustrative material, but it is possible to fathom some reasons, such as the practical, financial constraints on publishing illustrated natural histories that Chapter 1 has touched upon. Engraved images were costly, and thus rendered books too expensive for the average

---

173 Laurens Theodorus Gronovius to Arnout Vosmaer, 28 December 1751, Special Collections Department of Universiteitsbibliotheek Leiden (hereafter UBL), Leiden, BPL246, 51r.

174 Antoine Gouan to Linnaeus, 8 January 1760, UUB, L2656.

175 Jacob Theodor Klein, *Petri Artedi operum brevis recensio*, BL, Sloane MS 4020.

176 He later published his own work on fish: Jacob Theodoor Klein, *Historiae piscium naturalis* (Gdansk: Thomas Johann Schreiber, 1740–1749).

177 Banks's annotated copy of *Ichthyologia*, BL, 956c16.

178 Linnaeus' annotated copy of *Ichthyologia*, Linnean Society (herafter LS), London, BL.144. For more on this practice, see: Edwin Rose, "Specimens, Slips and Systems: Daniel Solander and the Classification of Nature at the World's First Public Museum, 1753–1768," *British Journal for the History of Science* 51, no. 2 (2018): 205–237.

student of natural history.[179] It might well be the case that he did not consider it worth the effort. Linnaeus' *Systema naturae*, for example, appeared virtually without images because he found that his system was best expressed linguistically rather than visually.[180] While admitting that illustrations "conveyed something to the unlearned", he also said that it was his intention to "try to express by words all features just as clearly – if not more clearly – as others with their splendid drawings."[181] Artedi's readers expressed similar sentiments. In another review of the *Ichthyologia*, published in the *Neue Zeitungen von gelehrten Sachen* in 1738, the reviewer stated that "[…] in his descriptions [Artedi] has shown such clarity that one can understand everything without illustrations, such that one is capable of placing them in their proper classes, a thing that nobody has managed before him."[182] Perhaps it was alien to Artedi's approach to produce illustrations, as these were only a poor substitute for the essence of the fish itself.

This is not to say that, for Artedi, illustrations did not merit attention. In his bibliographical overview of fish books, Artedi opined on the quality of images. In his *judicium* of the illustrations in Willughby's and Ray's *Historia piscium*, for example, he criticises the wood-cut images of Marcgraf for being particularly crude [*rudissimae*].[183] Those images taken from Rondelet and Belon had been executed slightly better, but were nonetheless underdeveloped. He was most pleased with those of Leonhard Baldner's drawings that had been copied into the book: he found them truly very handsome and, in all, most accurate [*vero nitidissimae et omnium accuratissimae*], although he did not specify what exactly made them so good.[184] Abstaining from using illustration in one's work does thus not necessarily indicate an aversion to the visual representation of nature. Charmantier has argued that the latter played a considerable role in Linnaeus' thinking; while he rarely included images in his work, he did include ample maps, diagrams and tables.[185] It was thus not that Linnaeus and Artedi were blind to the visual representation of nature, but rather that they did not

179 See: John L. Heller, "Linnaeus on Sumptuous Books," *Taxon* 25, no. 1 (1976): 33–52.

180 Dániel Margócsy, "'Refer to folio and number': Encyclopedias, the Exchange of Curiosities, and Practices of Identification before Linnaeus," *Journal of the History of Ideas* 71, no. 1 (2010): 83.

181 Linnaeus, *Genera plantarum* (Leiden: Conrad Wishoff, 1737), aph. 13, at 8; as translated in Müller-Wille and Reeds, "A Translation," 568.

182 "Petri Artedi," *Neue Zeitungen von gelehrten Sachen* 24, no. 1 (1738): 347–351, as translated in Pietsch and Aili, "Jacob Theodor Klein's Critique," 58.

183 *Icht.*, *Bibliotheca ichthyologia*, 52.

184 Ibid.

185 Charmantier, "Carl Linnaeus and the Visual Representation of Nature," 371.

consider illustrations indispensable for the identification of species, and at times even problematic.

Illustrations were also considered less advantageous than preserved specimens. To understand why, we turn to the aforementioned Laurens Theodorus Gronovius and his father, Johan Frederik (1690–1762). Civil servants of the city of Leiden, they were also avid collectors who knew Artedi's system intimately – as discussed earlier, they even seem to have been in the possession of one of the *Ichthyologia*'s original manuscripts. When Linnaeus resided in the Netherlands between 1735 and 1738, Gronovius the Elder had promptly recognised the merits of his classification scheme and helped to finance the first edition of the *Systema naturae*. Gronovius the Younger shared this appreciation, such that his father wrote of him: "he wants to describe all his pieces according to the manner of Linnaeus, […] and has already studied more than a hundred fishes, both indigenous and exogenous ones."[186] Even though Gronovius the Elder referred to it as 'the manner of Linnaeus', when it came to fish, 'the manner of Linnaeus' was largely that of Artedi – at least in the first few editions of the *Systema naturae*. That father and son Gronovius were familiar with Artedi's system becomes clear when perusing the *Museum ichtyologicum* (Leiden, 1754–1756). This catalogue of all specimens in their fish collection closely resembles the format of the *Ichthyologia*, following Artedi's sequence of orders, genera and species. It also makes frequent reference to Artedi's work: in describing a species of smelt, for example, the author added '*Arted. Gen.* 8' and '*Synon. p.* 21', so that the reader would know which genus this was in the *Ichthyologia*, where he might find a description of it, as well as a survey of the synonyms for this species.[187]

According to Margócsy, the need for classificatory repertories, like catalogues enumerating brief morphological descriptions of many species, was intimately linked to the increasing exchange of specimens. Plants, shells and insects were traded with particular avidity, on account of their modest size and weight, and the demand for ways with which collectors might clearly categorise them grew accordingly. For birds, fish and quadrupeds, there was less need for classificatory repertories because they were too expensive to ship in equally large numbers.[188] There is certainly truth to that pronouncement: whether, as

186 As translated by me from Dutch quotation in Bert Sliggers, *De Verzamelwoede van Martinus van Marum (1750–1837) en de Ouderdom van de Aarde: Herkomst en Functie van het Paleontologisch en Mineralogisch Kabinet van het Teylers Museum* (PhD diss., Leiden University, 2017), 156.

187 Laurens Theodorus Gronovius, *Museum ichthyologicum, sistens piscium indigenorum et quorumdam exoticorum* (Leiden: Theodoor Haak, 1754–1756), 18.

188 Margócsy, *Commercial Visions*, 32.

we saw, in the case of larger fish, specimens were stuffed with hay, or, with smaller fish, submerged in spirits, specimens prepared in either way would take up considerable space in the hold of a ship or on a carriage. But, as we will now see, this was not necessarily always the case. Naturalists did have ways of preserving fish in a more compact fashion.

In the sixteenth century, Salviani had wondered whether it would be possible "to find a way to preserve dried fish in their own shape, like one does with these herbs."[189] The herbarium, in which plants were fixed to pages with glue or thread, was a favoured method of botanical preservation at the time.[190] While he did not offer a new name for such an extended use of this cherished tradition, what Salviani proposed was what we will from here on refer to as an 'ichthyarium': an herbarium that contains the skins of fish instead of flowers and plants.[191] Despite Salviani's efforts to glue fishes to paper in the sixteenth century, it seems that his particular style of ichthyarium did not catch on widely, even if archives contain evidence of occasional use: there is, for example, an instance of a fish skin glued on paper among Sloane's papers.[192] The paper ichthyarium became a widespread method of circulating fish specimens only after Gronovius the Elder wrote out his method for preserving fish on paper in the early 1740s.

In 1742, Gronovius the Elder communicated his method to Peter Collinson (1694–1768), who read the letter aloud during a meeting of the Royal Society and had it published in the *Philosophical Transactions*.[193] To enact the procedure successfully, one needed a pair of scissors with very fine blades and sharp points, small wooden plates (preferably of the lime tree), a very fine needle, slips of parchments as large as the fishes, and finally some very small pins. With all these materials in place, the preservation process could commence. The method, which Gronovius laid out in his letter step-by-step, entailed

189 Findlen and Toledano, "Materials of Natural History," 158.

190 The history of the herbarium is described in detail by Ogilvie, *Science of Describing*, 165–174.

191 The term 'fish herbarium' would be incongruous, as the term herbarium refers to herbs (from the Latin *herba*). The Greek word 'ichtus' or ἰχθύς refers to fish, and ἰχθύα can also refer to the skin of a fish. *Logeion*, s.v. ἰχθύς and ἰχθύα.

192 Fish skin pasted on paper, BL, Add MS 5267 (part of Sloane's collection), f99r; it is dated 1732. I thank Felicity Roberts for bringing it to my attention.

193 J.F. Gronovius, "A Method of preparing Specimens of Fish, by drying their Skins, as practised by John Frid. Gronovius M.D. in Leyden," *Philosophical Transactions of the Royal Society* 42, no. 463 (1742): 57–58. For the practice of reading letters aloud around this time, see: Aileen Fyfe and Noah Moxham, "Making Public Ahead of Print: Meetings and Publications at the Royal Society, 1752–1892," *Notes and Records of the Royal Society of London* 70, no. 4. (2016): 361–379.

gently cutting a fish open with a pair of scissors, removing all of its intestines, patting it dry with a linen cloth, exposing it to the sun (in summer) or to the hearth (in winter) for further drying. The skin could then easily be separated from the flesh and was to be put between papers and pressed flat. Gronovius advised the application of a fresh sheet of parchment after two hours, "as a sort of glutinous Matter, in pressing, is always forced out from betwixt the Scales and the Skin" which might cause the fish to stick to the paper.[194] Only a few materials and a little time were needed, as Gronovius stated that "in the Space of 24 Hours, the Fish is prepared."[195] Even if the parchment were changed to absorb excess quantities of the glutinous matter that emanated from the remains of the fish, enough of the substance might have well remained to give the specimen a self-adhesive quality.[196]

In 1751, Gronovius sent a package to Pennsylvania that included "a few specimens of dryed fishes, to be kept as plants in an herbarius; the great misfortune is, that the colour perish, else it shows a good way to find out the characters 1. by number of the bones in the gill flap [...]; 2. by the number and position of the fins, and the bones in them. 3. by the course of the linea lateralis running in each fish from the back part of the head to the tail."[197] As Gronovius' letter indicates, the method preserved precisely those characteristics that were of most pertinence for classification according to Artedi's system. One of the things that unfortunately could not be salvaged was colour; on the goldfish, Gronovius the Younger remarked that "the colour in life is gold or silver, shining and most radiating, which in death gradually disappears and whitens."[198] The colour thus died together with the fish. Upon receiving a species of sandfish from the Cape of Good Hope, which he submitted was 'entirely new' [*plane novus*], he declared he could "barely guess the colour from this dried specimen."[199] The specimen that he referred to, and which he called the *Gonorynchus*, is

---

194 Gronovius, "A Method of preparing", 57–58.

195 Ibid.

196 Fish glue has historically been used for a wide range of artistic practices, see: Tatyana Petukhova, "A History of Fish Glue as an Artist's Material: Applications in Paper and Parchment Artifacts," *The Book and Paper Group Annual* 19 (2000): 111–114.

197 J.F. Gronovius to John Bartram, 26 June 1751, as reproduced in *The Correspondence of John Bartram 1734–1777*, eds. Edmund Berkeley Jr. and Dorothy Smith Berkeley (Gainesville: University Press of Florida, 1992), 330.

198 "Color in vivus aureus vel argenteus, splendidus & fulgentissimus, qui in mortuo sensim perit & albescit." Laurens Theodorus Gronovius, *Zoophylacium Gronovianum: exhibens animalia quadrupeda, amphibia, pisces, insecta, vermes, mollusca, testacea, et zoophyta* (Leiden: Theodoor Haak, 1781), 109.

199 "Colorem ex siccato specimine divinare haud potui [...]". Ibid., 55.

FIGURE 13 Specimens of *Gonorkynchus greyi* (above) and *Misgurnus fossilis* (below). Gronovius Fish Collection, BMNH 1853.11.12.120, Trustees of the Natural History Museum in London
PHOTO BY LUCIE GOODAYLE

still among the Gronovius Fish Collection at the Natural History Museum in London (Figure 13).[200]

Certain matters were definitely lost in preservation. As has already been discussed in Chapter 2, illustrations could convey liveliness in ways that dead, preserved specimens could not. The addition of colour to illustrations was significant, as this was one of the things that disappeared after death. Because colour carried no value for classification, however, Gronovius the Elder found there was no need to take the trouble. Furthermore, illustrations were mediations, and often did not relay the characteristics that the classifying naturalist desired to know – which might be because the artist was not aware of these marks, or because these could not easily be conveyed on paper. Gronovius stated, for example, that "Mr. Catesby hath indeed painted the American fishes very well. But I wish to know of all them fishes, how many bones there are in the gill flap, which is sufficient to determine the genus together with the

200 Specimen of *Gonorkynchus greyi* [*sic*], currently accepted species name: *Gonorynchus greyi*, Natural History Museum (hereafter NHM), London, Gronovius Fish Collection, BMNH 1853.11.12.120. The NHM keeps the ichthyarium of Gronovius, consisting of loose pages, in several dozen boxes. The fish retain some of their original odour.

number of fins."[201] He referred to the *Natural History of Carolina, Florida and the Bahama Islands* (London, 1731) by the English naturalist Mark Catesby (1683–1749), a work that contemporaries praised for its perfectly executed watercolours.[202] Gronovius's remark underscores the earlier mentioned belief that illustrations, however pretty or skilfully done, were superfluous to the core business of natural history.

Quite a few did take up the method proposed by Gronovius the Elder. Among them were the French naturalists Michel Adanson (1727–1806) and Philibert de Commerson (1727–1773), as well as Linnaeus.[203] In his autobiography, Linnaeus wrote how in his collection "were innumerable stones, in his herbarium and Garden innumerable plants, in his cabinet innumerable insects that he had assembled and pinned, in his cupboards innumerable fish glued on paper as if they were plants [...]."[204] The fact that fish specimens were pressed flat enough to be inserted into a letter wrapper facilitated their circulation between various countries and continents in large quantities and at relatively low cost, and with less chance of being damaged en route.[205] However, the method also created challenges. What to do with species of flat fish, such as flounders or rays? Or species that might not comfortably on a piece of paper altogether, like the European eel? Barlow Robles has shown how, while the preservation method was key in building a comprehensive and universal system, fish often resisted their flattening.[206]

---

201 Johan Frederik Gronovius to Bartram, 6 December 1745, as reproduced in Berkeley and Berkeley, *The Correspondence of John Bartram*, 265.

202 Amy R.W. Meyers and Margaret Beck Pritchard, "Introduction: Toward an Understanding of Catesby," in *Empire's Nature: Mark Catesby's New World Vision*, eds. Amy R.W. Meyers and Margaret Beck Pritchard (Chapel Hill: University of North Carolina Press, 1998), 5.

203 Amandine Péquignot, "Une peau entre deux feuilles, l'usage de l'herbier en taxidermie aux XVIIIe et XIXe siècles en France," *Revue d'histoire des sciences* 59, no. 1 (2006): 131–132. This article explains how the skins of birds as well as fishes were pressed between pages in herbarium fashion.

204 Wilfrid Blunt, *The Compleat Naturalist: A Life of Linnaeus* (London: Collins, 1971), 151. Most of the fishes in his collection are now in the Linnean Society in London and the Gustavianum, the universitetsmuseum of Uppsala, and many are considered type specimens. See also: Alwhyne Wheeler, "The Linnaean Fish Collection in the Linnean Society of London," *Zoological Journal of the Linnean Society* 84, no. 1 (1985): 1–76; Alwhyne Wheeler, "The Linnaean Fish Collection in the Zoological Museum of the University of Uppsala," *Zoological Journal of the Linnean Society* 103, no. 2 (1991): 145–195.

205 Whitney Barlow Robles, "Flatness," in *The Philosophy Chamber: Art and Science in Harvard's Teaching Cabinet, 1766–1820*, ed. Ethan W. Lasser (New Haven: Yale University Press, 2017), 196.

206 Barlow Robles, *Curious Species*, 140, 155.

While the spread of the ichthyarium as preservation practice was partly induced by practical circumstances, it was also welcomed because it adhered to the epistemological requirements of classification. As we saw, the essential characteristics, such as the fins and their rays and the bones in the gill flap once preserved in this manner, could be pressed into the service of the inquisitive naturalist. Using this method, naturalists could receive the actual specimens rather than mediated illustrations, and therefore did not need to rely on artists whom they had never met and whose skills they therefore had little way to assess. For adherents of Artedi's classification system, this method was thus not only a practical way to preserve fish given the constraints of money and material. As 'incarnations of themselves' (to borrow a phrase by Daniela Bleichmar)[207] fish prepared in this manner were considered by naturalists epistemologically better suited than images to base their classifications on. In other words, Gronovius here made an epistemic choice – he found objects preferable to images (which could be wrong) as a basis of classification.

Charmantier and Staffan Müller-Wille have argued for Linnaeus' works that they were useful to naturalists in the complex process of comparing specimen to specimen, distinguishing one species from the other, and in drawing up their own species diagnoses.[208] It made it easier to count and categorise species, and to assign names and numbers to them.[209] Even though Artedi, unlike Linnaeus, did not present his readers with binomial names consisting of one word for the genus and one word for the species, his *Ichthyologia* did offer a clear overview of both genera and species, their characteristics enumerated in the form of numbered lists. It was praised by naturalists for its precise and distinct method of classification. The clear species descriptions meant that illustrations were not necessary. Because the ichthyarium method made popular by Gronovius allowed for the dispatching of fish specimens from all corners of the world in an economical fashion while – ideally – retaining those characteristics that were important in classification, it was privileged as a form of preservation. It allowed naturalists to count these characteristics for themselves.

---

207 Bleichmar, *Visible Empire*, 63.

208 Isabelle Charmantier and Staffan Müller-Wille, "Carl Linnaeus's Botanical Paper Slips (1767–1773)," *Intellectual History Review* 24, no. 2 (2014): 227.

209 Staffan Müller-Wille, "Names and Numbers: 'Data' in Classical Natural History, 1758–1859," *Osiris* 32 (2017): 120–126.

5 Conclusion

What makes Artedi's book a compelling source is its detailed discussion of the principles that underpin his classification, and how in it he equated his method with the field of 'ichthyology.' His work offered rules for how ichthyology should be done – and by whom. As such, it reads like a handbook or code of practice. We find in it the first explicit articulation of what 'ichthyology' is (or, at least, what he thought it ought to be) as well of the first definition of its prime practitioner: the 'ichthyologist'. As Artedi explained, an ichthyologist adheres to the three-tiered hierarchical system that he developed. This elevated the ichthyologist as an authorative knower of fish, quite apart from those who examined fish in an 'amethodic' manner, such as those who wrote about fishery. The ichthyologist, furthermore, made true distinctions, whereas those distinctions made by fishermen tended to be false and frivolous. In rejecting the knowledge of these artisanal or practical communities in this way, foregoing his own reliance on their observations both in the field and as recorded in the books of natural history he used as sources, Artedi marked out the ichthyologist as its own separate category of expert.

Artedi's book was not only an attempt to classify fish, but also to classify past and present knowledge about fish and, by extension, people who studied and worked with fish. Moreover, it was an attempt to establish ichthyology as a science; to define its object, and to unfold its epistemology. In what would be the last letter he wrote to his relatives, Artedi spoke confidently of his mastery of ichthyology, asserting that "in the Dutch Republic, there is no one who understands zoology anyway, and where Ichthyology is concerned, I admit neither there nor elsewhere anyone for my master."[210] It reveals how Artedi considered himself as a specialised naturalist with an authoritative command of the natural history of fish. We will now recall the opening story of this chapter, where Artedi longed to confer with an ichthyologist – not any, but rather a 'true' [*vero*] one. Similar qualifications were made by Linnaeus, who referred to botanists as 'sound' [*sanis*].[211] So what did it take to be a sound botanist or true ichthyologist?

For Artedi, the answer to this question was his own method. In developing it, he seems to have had mind how it would distinguish him from other

---

210 "Här i Holland är ingen särdeles såm förstår sig på Zoologien, och hwad Ichthyologien angår, så ärkiänner jag hwarken där eller annorstädes någor för min maitre [...]" KBS, MS X1002.

211 Linnaeus, *Genera plantarum*, 3.

naturalists and serve to establish his name. We see this, for example, in his interactions with Sloane, to whom he gave a manuscript displaying his classification method for fish, and with Seba, who hired Artedi to describe his species according to it. As his life was cut short before he had fully prepared his work for publication, we cannot know to whom his dedicatory letters would have been addressed. Earlier studies of the *Ichthyologia* have overlooked the social context in which the work was produced; so far, attention has always been directed to the classificatory contents of the book itself rather than its style of writing, layout, or structure. This chapter has shown the rhetorical effects of such decisions. In unfolding his new system, Artedi grafted his book onto traditional models, drawing on humanist and scholastic models of organising knowledge, such as the genre of the *literaria* and the format of the *disputatio*, as well as by selecting Greek and Latin as the only languages acceptable for the names of genera and species. By doing so, he anchored his novel approach to long-standing scholarly traditions of organising and presenting knowledge.

His method was novel indeed. As we saw in Chapter 1, Renaissance naturalists had grouped species together based on a wide range of characteristics, for which they consulted a diversity of sources including literary ones. Willughby and Ray had proposed a strict focus on external characteristics, and had used these to construct taxonomical groups, albeit loosely. While they eschewed humanistic learning, their species descriptions nonetheless incorporated various anecdotes, proverbs and matters relating to fishery. For Artedi, it was not relevant whether a fish dwelled in fresh or salt water, with which letter of the alphabet its name started, or how it figured in trade. Artedi drew up strict definitions for taxonomical ranks on all levels, arranging his descriptions in list format. What earlier authors had written on a certain species only mattered in the extent to which they corresponded with his own descriptions. Artedi's method imposed abstraction by including only those characteristics that were pertinent to classification, and leaving out the rest. The innovative character of this system cannot be overstated. In the process, the study of fish became an increasingly visual and almost quantative pursuit, as closely observing species and carefully counting their physical characteristics became one of its hallmarks. Although one would expect that this would make images a vital part of natural historical study, this was not necessarily the case for Artedi and his adepts. Images were not only expensive to produce, they might also be unreliable if artists were either unaware or unable to capture these aforementioned characteristics. A preserved specimen with its fins, teeth, and other parts intact would better perform its function for classification.

The model that Artedi developed for classifying fish meant that one could take any fish, known or unknown, and through close observation navigate one's way through the correct order, via the genus, and arrive at the correct species. With this system, Artedi widened the realm of possible observers, as a fish could now be assigned its proper place in Creation simply by counting fins and checking other characteristics. He also narrowed it down, however, by deeming only particular kinds of knowledge pertinent for the field that he defined as ichthyology. As we have seen in the previous chapter, early modern natural history took place in a composite social space. In excluding artisanal and vernacular knowledge, Artedi whittled away this longstanding practical knowledge from the body of natural historical knowledge about fish, lending his own methodical principles more authority in the process. He thus narrowed the kind of knowledge that was of possible import to learned inquiries into fish.

Linnaeus, similarly, placed those who worked with nature into various categories and assigned different values to them. For him, observers of plants were interchangeable as long they had mastered a standardised set of skills, namely knowing what to look for (such as the numbers of stamen and pistils).[212] This only held, however, for the practical work of recognising and defining species of plants; the more theoretical and abstract work of drawing up a scheme of genera, orders, and classes would have to be done by those with specialist knowledge: the botanist.[213] Harriet Ritvo has argued that after the introduction of classification systems, "taxonomy itself became a characteristic of the highest taxonomical significance."[214] In other words, an understanding of classification systems became to be seen as an indication of one's proficiency in natural history, as well as one's ability to reduce living creatures to some underlying pattern.

Linnaeus' success was to no small degree due to the 'templates for communal annotation' that his taxonomic works offered.[215] These templates created a community that worked towards the collective goal to complete the system by fitting ever more species into it based on observations in the field, the museum or the library.[216] As a result, the system that Linnaeus had designed to contain and instill order into information about species contributed to the propagation of this kind of information.[217] While barely registering by comparison with the

212 Cooper, *Inventing the Indigenous*, 171.

213 Nickelsen, *Draughtsmen*, 74.

214 Harriet Ritvo, "New Presbyter or Old Priest? Reconsidering Zoological Taxonomy in Britain, 1750–1840," *History of the Human Sciences* 3, no. 2 (1990): 260.

215 Staffan Müller-Wille, "Linnaean Paper Tools," 210.

216 Staffan Müller-Wille and Sara T. Scharf, "Indexing Nature: Carl Linnaeus (1707–1778) and His Fact-Gathering Strategies," *Svenska Linnesällskapets Årsskrift* 94 (2011): 57.

217 Isabelle Charmantier and Staffan Müller-Wille, "Natural History and Information Overload: The Case of Linnaeus," *Studies in History and Philosophy of Science Part C: Studies in History and Philosophy of Biological and Biomedical Sciences* 43, no. 1 (2012): 4.

copious translations, adaptations and editions that Linnaeus' work inspired, Artedi's work was also revisited. In 1789, for instance, a revised edition of his book was issued in the German states. Johann Julius Walbaum (1724–1799) saw fit to publish the *Petri Artedi Renovati,* in which he amended Artedi's species descriptions, and added species and genera that had been described after the publication of the book.[218] Like the system of Linnaeus, that of Artedi was flexible enough to allow for adaptations by later naturalists while keeping its basic structure intact.[219] The following chapter will look at one such naturalist, Marcus Élieser Bloch, who drew on this classification system, but took other approaches to both description and illustration.

218 Johann Julius Walbaum, *Petri Artedi Renovati* (Greifswald: Anton Ferdinand Röse, 1789).

219 Bettina Dietz, "Linnaeus' Restless System: Translation as Textual Engineering in Eighteenth-Century Botany," *Annals of Science* 73, no. 2 (2014): 155.

CHAPTER 4

Swimming on the Page: Illustration and Image in Marcus Élieser Bloch's *Allgemeine Naturgeschichte der Fische*

How can one capture a fish on paper? This question was often on the mind of the Jewish-German naturalist Marcus Élieser Bloch. His ambitious and costly book series, the *Allgemeine Naturgeschichte der Fische* – twelve lavishly illustrated volumes on fish published from 1782 to 1795 – was the result of decades of research into these creatures and contained over four hundred hand-coloured engravings. These coloured illustrations were the defining feature of the work. In 1794, the missionary Christoph Samuel John (1747–1813), who had contributed a considerable number of specimens from the Coast of Coromandel where he was stationed, wrote to Bloch describing admiration Bloch's series of fish books elicited in all those who saw them. Published in the *Berlinische Monatschrift* for all of the periodical's enlightened audience to read, John's letter went on to explain how the black fishermen who supplied him with specimens for him "were astounded by how lively their fishes swam about on the page."[1] Whether this exchange actually took place, this was an exaggerated account of it, or whether John wanted to present Bloch with a flattering fiction, the terms that he used are significant. After all, Bloch had taken great pains to ensure that each and every illustration in the book series portrayed the fish species as they must have looked before death and subsequent decay. To replicate the glistening of scales, his artists sometimes even used paints that contained from actual silver and gold.

While it was common practice for authors to copy illustrations, the majority of the 432 plates in Bloch's work had been designed anew by artists. The illustration of a species of catfish that John had sent him, which Bloch had named the bimaculated silure (Figure 14), shows how the engravings were carefully hand-coloured, delicately depicting the species' various hues.[2] By ingeniously combining techniques for engraving and colourisation, Bloch crafted his

1 "[...] und meine schwarzen Fischer können nicht genug erstaunen, wie ihre Fische itzt [jetzt] auf dem Papier so lebendig herumschwimmen." Christoph Samuel John, "Einige Nachrichten von der Küste Koromandel. Auszug eines Schreibens des Hrn Missionarius C.S. John an Hrn D. Bloch in Berlin," *Berlinische Monatschrift* 24 (1794): 351.

2 The Latin name Bloch gave this fish was *Silurus bimaculatus.*; its currently accepted species name is *Ompok bimaculatus*.

DOI:10.1163/9789004549500_006

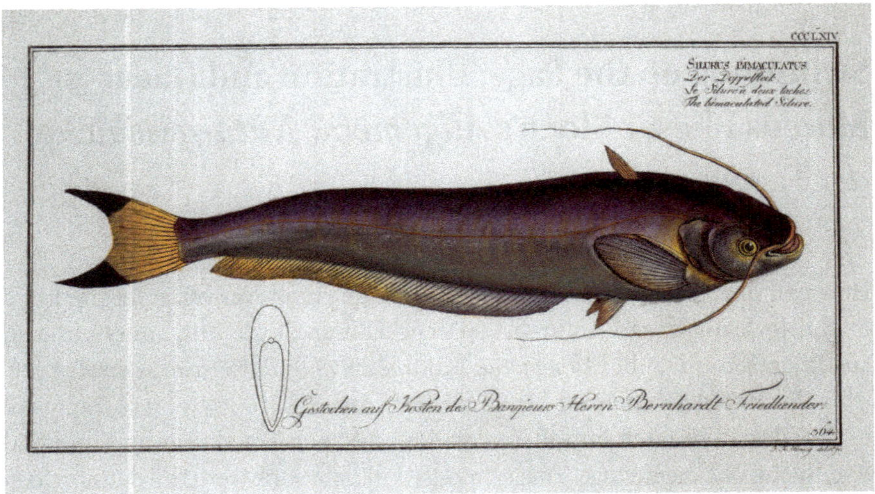

FIGURE 14 Engraving of *Silurus bimaculatus*, Johann Friedrich Hennig, vol. 12 (Berlin: J. Morino, 1795), plate CCCLXIV
NEW YORK PUBLIC LIBRARY

illustrations to function as 'epistemic images' which distilled the characteristics of fishes that he considered essential so that the image could serve as a substitute for the depicted species.[3] In this sense, his books are rather different to those of Linnaeus and Artedi, which appeared virtually image-free. While Bloch continued the classificatory approach that the Swedish naturalists had promulgated in their works, he could not imagine doing without illustrations. In fact, in his introduction to the book series he declared that he was only including those fishes for which he was able to offer drawings done after nature.[4] Believing that illustrations allowed for the most accurate capture of fish, Bloch took pride in sparing neither effort nor expenditure in giving his volumes "the highest degree of perfection."[5]

Looking beyond the beautiful images, the books bring together a rather surprising constellation of people. As the list of subscribers to the volumes shows, his books were of interest to a Luxembourgian fishing guild as well as to royals and nobles, government officials, bankers, apothecaries, physicians, preachers,

3 For the notion of 'epistemic images', see: Lorraine Daston, "Epistemic Images," in Alina Payne (ed.) *Vision and Its Instruments: Art, Science, and Technology in Early Modern Europe* (Pennsylvania State University Press: Philadelphia, 2015), 13–35.

4 "Ich werde mich indessen nur auf solche Fische einschränken, von welchen ich nach der Natur gemalte Zeichnungen zu lieferen im Stande bin [...]." *Allg. Nat. der Fische*, vol. 1, sig. Ar.

5 "[...] da ich weder Mühe noch Kosten gesparet, um demselben den möglichsten Grad der Vollkommenheit zu geben." *Allg. Nat. der Fische*, vol. 3 (Berlin: Realschule, 1784), n.p.

and booksellers.[6] It is an example of how the European audience for "visually sumptuous and winningly designed volumes", was growing in the eighteenth century.[7] Yet it is also indicative of how Bloch, a Jewish naturalist who was not granted full legal equality with the other inhabitants of his city of Berlin, found a way to fully integrate in learned circles. The work was, furthermore, produced within a colonial context. Even though the early modern German states did not maintain a consistent presence outside of Europe, Bloch benefited from colonial infrastructures in expanding his collection. John, who was the main contributor of foreign species to Bloch's project, was a Pietist missionary based in South India who relied on individuals in and around the mission to furnish him with specimens. The purpose of this chapter is to show how Bloch's fish series, which promoted these coloured engravings as its unique selling point, helped him to negotiate intricate social structures, and, furthermore, that these engravings were instrumental in the creation of his image as a naturalist.

The chapter first introduces Bloch and his collection, in order to embed him in the wider community of late eighteenth-century *Naturforscher*. The second section considers how his series of fish books straddled various geographical boundaries and catered to varied audiences. The third section traces how those specimens that hailed from Europe's colonial outposts had reached the shelves of Bloch's cabinet in the first place, focusing on the most important contributor of colonial specimens, the abovementioned German missionary John, who enlisted local assistants to collect fish on the Coromandel Coast of South India. The fourth and last part of the chapter analyses the way in which Bloch's fish collection was preserved on paper, with particular emphasis on the way in which the illustrations were designed and executed and how they functioned as an act of preservation. Ultimately, it argues that the manner in which Bloch presented his collection on paper served to make him an authority on the fishes of the world without requiring him to travel outside of his Berlin.

1 Collections, Identities, and Reputations

It was in the early 1770s that Bloch had started to assemble an impressive collection in his home on the Spandauerstraße in Berlin, where the more prosperous Jewish inhabitants of the city lived.[8] In his guide to notable places

6 *Allg. Nat. der Fische*, vol. 1, n.p.

7 Schmidt, *Inventing Exoticism*, 18.

8 Steven M. Lowenstein, *The Berlin Jewish Community: Enlightenment, Family, and Crisis, 1770–1830* (Oxford: Oxford University Press, 1994), 16. Bloch's collection is mentioned in the

in the area of Berlin and Potsdam, the printer Christoph Friedrich Nicolai (1733–1811) described it as a "well-ordered collection of *naturalia* contained in eight glass cabinets and five drawered chests", which accommodated a broad range of natural productions. Amongst the many items he mentioned were fossils and polished stones, wood samples, endemic and foreign birds with their nests, reptiles ("many beautiful snakes"), several boxfishes and a flying fish, rare shells, large Surinam beetles, and even human remains, of which he found "an embryo conceived by an European and a *Mohrin*" most noteworthy.[9] This description demonstrates both the breadth of the collection and the colonial context which shaped it. Bloch's cabinet was not in itself unique; Nicolai described 27 other collections in Berlin alone.[10] In eighteenth-century Europe, increasing numbers of merchants, physicians, professors and others amassed collections of rare and valuable natural and cultural objects in their households.[11] Bloch's collection, however, was held in particularly high esteem. Even the great Wilhelm von Humboldt spoke of the 'Bloch'sche Cabinet', writing to Johann Wolfgang von Goethe that he heard it held all kinds of rarities.[12]

What made this collection truly stand out, however, was its vast amount of fish. By the time of his death in 1799, Bloch's collection of fish had grown to encompass almost 1,400 specimens.[13] Over 1,000 of these floated in glass jars full of alcohol, while others were stuffed with hay or existed as preserved skins, some of which were mounted on wooden models. Why this focus on fish? It is a question the naturalist answered on the very first page of the preface to the very first volume of the *Allgemeine Naturgeschichte der Fische*.[14] In this preface, Bloch explained how, given that he was spending much of his leisure time

---

Tagebuch entry of 17 August 1773 of the Gesellschaft Naturforschender Freunde. MfN, ZMB, GNF, S. Bloch, TB 1.

9 Christoph Friedrich Nicolai, *Beschreibung der Königlichen Residenzstädte Berlin und Potsdam* vol. 2 (Berlin: Friedrich Nicolai, 1779), 599–601. An abbreviated, paraphrased translation can be found in Antoine-Joseph Dézallier d'Argenville, *La conchyliologie, ou, Histoire naturelle des coquilles de mer, d'eau douce, terrestres et fossiles* (Paris: Guillaume de Bure fils aîné, 1780), 828.

10 Nicolai, *Beschreibung*, 598–609.

11 For a recent overview of types of collections in eighteenth-century Europe, see: Eva Dolezel, Rainer Godel, Andreas Peča and Holger Zaunstöck, eds., *Ordnen – Vernetzen – Vermitteln. Kunst- und Naturalienkammern der Frühen Neuzeit als Lehr- und Lernorte* (Stuttgart: Wissenschaftliche Verlagsgeselschaft, 2018).

12 Wilhelm von Humboldt to Johann Wolfgang von Goethe, 22 August 1795, *Goethes Briefwechsel mit den Gebrüdern Humboldt (1767–1832)*, ed. F.Th. Bratranek (Leipzig: F.A. Brockhaus, 1876), 7.

13 Eva Dolezel, "Lehrreiche Unterhalting oder Wissenschaftliche Hülfsmittel? Die Berliner Kunstkammer um 1800. Eine Sammlung am Schnittpunkt Zweier Musealer Konzepte," *Jahrbuch der Berliner Museen* 46 (2004): 151–152.

14 *Allg. Nat. der Fische*, vol. 1, sig. *2r.

perusing natural history, a friend sent him a salmon from the lake of Miedwie, in the Province of Pomerania bordering on the east of Brandenburg. On consulting his copy of Linnaeus' *Systema naturae* he found that not only did this particular species of salmon go unmentioned, but so too did many other fish common in the German states.[15] Germanic Europe may have been, in the words of Kathryn Olesko, "culturally diverse, politically fragmented and geographically fluid", Bloch saw it as an area unified enough to be charted through its fish.[16]

Just as Willughby and Ray had done a century before him, Bloch obtained specimens at marketplaces and harbours. While his initial intention was to chart only German fishes, his collection soon expanded to what he called *fremde Fische* [foreign fish], that is to say fish not native to the German states.[17] Bloch boasted that he received fish from all corners of the world through his network consisting of government officials, physicians and missionaries. His collection thus grew to encompass specimens that originated in Scandinavia, Greenland, the North Atlantic, the Mediterranean, Africa's west coast, the Caribbean, Surinam and Brazil, North America and South India.[18] A further section of this chapter will explore the various individuals who contributed the specimens from the latter region that ended up on Bloch's shelves.

As various historians have shown, collections were tied closely to identities. They were about making and displaying connections: presenting objects from faraway places allowed the collector to impress viewers with their global reach.[19] In his biography of Hans Sloane, James Delbourgo showed how as the "collector of collectors", Sloane created a central position for himself in the learned world of the late seventeenth and early eighteenth centuries.[20] Collectors did not merely demonstrated their connections, however, but also their skill and taste. As Emma Spary has noted, collections were thought to represent the collector's personal character, so that making an orderly and aesthetically pleasing cabinet was a good way of establishing one's reputation.[21]

15 He does not mention which edition.

16 Kahtryn Olesko, "Germany," in Hugh Richard Slotten, Ronald L. Numbers and David N. Livingstone (eds.) *The Cambridge History of Science: Modern Science in National, Transnational, and Global Context* (Cambridge: Cambridge University Press, 2020) 23.

17 *Allg. Nat. der Fische*, vol. 3, n.p.

18 Paepke, *Bloch's Fish Collection*, 27.

19 Daniela Bleichmar and Peter C. Mancall, "Introduction," in *Collecting Across Cultures: Material Exchanges in the Early Modern Atlantic World*, eds. Daniela Bleichmar and Peter C. Mancall (Philadelphia: University of Pennsylvania Press, 2013), 1.

20 Delbourgo, *Collecting the World*, 202.

21 Emma Spary, "The Naturalist Collecting Community in Paris, 1760–1789: A Preliminary Survey," in Dolezel, Godel, Peča and Zaunstöck, *Ordnen – Vernetzen – Vermitteln*, 310.

Such factors may well have played a role in Bloch's decision to build a collection. As a Jew, Bloch was perceived as a relative outsider by the learned community. This was not, of course, because Jewish thinkers did not have a profound influence on early modern natural philosophy – quite the contrary. Compared to other naturalists, however, Bloch's entry to the world of learning was conditional, as we will see, both in a judicial and a cultural sense. We will now look at his emergence as a collector and naturalist in more detail.

Bloch was born in 1723 in Ansbach, Bavaria, and grew up in a rather impoverished, orthodox Jewish household.[22] His upbringing was traditional; he was taught Hebrew, for example, but probably not how to read in the Latin alphabet, as this was not encouraged in orthodox circles.[23] Sometime around 1743 he travelled to Hamburg, where he had obtained a position as teacher to the son of a Jewish surgeon. This is where he learned German and Latin, and where he acquired some basic medical knowledge. He then travelled to relatives in Berlin in order that he might study anatomy. When Bloch decided to pursue a doctorate in Medicine in 1760, he moved to Frankfurt am Oder; this city's university was the only one in Prussia that admitted Jews.[24] Having received his degree in 1762, Bloch settled in Berlin where he set up practice as a physician, married, and involved himself in the city's intellectual life.

Soon after arriving in Berlin, Bloch became involved in the Haskalah, also known as the Jewish Enlightenment, which was spurred by a group of intellectuals that had moved to the city due to its growing reputation as a centre of scholarship.[25] As home to the centre of Prussian government, the royal court, and a military garrison as well as being an emerging commercial hub, Berlin had been growing rapidly in the first half of the eighteenth century.[26] Among its varied population were migrants from different religious backgrounds.[27] Although the city was relatively open to newcomers, Jews, as elsewhere in Prussia, did not have full legal equality. The government curbed their rights to marry, buy property, found businesses, or attend university.[28] The figureheads

22 The most detailed biographical sketch of Bloch has been written by Christine Karrer, who has pieced it together from a variety of sources, including the letters and accounts of some of Bloch's contemporaries. Karrer, 130–132.

23 Lowenstein, *The Berlin Jewish Community*, 189; Johannes Müller, "Distance, Geography, and Anecdote in M.E. Bloch's Natural History of Fishes." In Smith and Egmond, *Ichthyology in Context*, 614.

24 Karrer, "Marcus Elieser Bloch," 132.

25 Lowenstein, *The Berlin Jewish Community*, 34, 49. On the Haskalah, see also Shmuel Feiner, *The Jewish Eighteenth Century: A European Biography, 1700–1750* (Bloomington: Indiana University Press, 2020).

26 Lowenstein, *The Berlin Jewish Community*, 4.

27 For example, French Huguenots and Austrian Protestants; ibid., 19.

28 Ibid., 13.

of the Haskalah worked towards reforms to advance the legal status of the Jewish community in the city. Some of these, such as the philosopher Moses Mendelssohn (1729–1786) and the banker David Friedländer (1750–1834) were close friends with Bloch.[29] Bloch himself was one of the founders of Berlin's Jewish hospital.[30] He also dedicated the tenth volume of his series of fish books to the heir presumptive, Frederick VI of Denmark (1768–1839), in acknowledgement of a decree that ensured the rights of Bloch's 'suppressed brethren' by granting Jewish pupils access to apprenticeships with craftsmen.[31] It is the only instance in which Bloch explicitly alludes to his Jewish background in his books, and yet it is clear that this emancipation effort was important to him.

Following the death of his first wife in 1769, Bloch wedded the affluent Cheile Ephraim (1757–1780) in 1774, a marriage which might well have facilitated the expansion of his collection.[32] His first publication, a book of medical observations, appeared in that same year.[33] In the following decades he established himself as prolific author of natural history, spending most of his time on the study of fish.[34] He collected species of fish, scoured the natural historical literature to see if they had already been described, and classified them according to Linnaean principles if that had not yet been done. The first volume of his fish series was published in 1782. From this moment on, his collection and his series of fish books nourished, influenced and reinforced one another. In the prefaces to the published volumes, Bloch solicited ever more fish specimens and drawings from his readers. They plainly obliged, as over the years in which his fish series was published, he continued to receive a good supply of specimens, which he in turn converted into descriptions and depictions for his series. Even though the development of the collection and the series were closely connected, we will first turn to the formation of the former, and only subsequently discuss the latter.

Bloch moved not only within the circle of the Jewish enlightenment, but also sought out others who shared his fascination for the study of nature.

---

29 Shmuel Feiner, *The Jewish Enlightenment* (Philadelphia: University of Pennsylvania Press, 2002), 117, 315.

30 Karrer, "Marcus Elieser Bloch," 135.

31 Marcus Élieser Bloch, *Allgemeine Naturgeschichte der Fische*, vol. 10 (Berlin: J. Morino, 1793), vi. See also Martin Schwartz Lauzen, *Jews and Christians in Denmark: From the Middle Ages to Recent Times* (Leiden: Brill, 2015), 89–124.

32 Karrer, "Marcus Elieser Bloch," 134.

33 Marcus Élieser Bloch, *Medicinische Bemerkungen, Nebst einer Abhandlung vom Pyrmonter Augenbrunnen* (Berlin: Christian Friedrich Himburg, 1774).

34 Besides publishing his fish series, Bloch wrote, among other things, on opal, tortoises and bladder worms in the *Beschäftigungen* of the *Gesellschaft Naturforschender Freunde*. His *Systema ichthyologiae* was published posthumously by Johann Gottlob Schneider (1750–1822) in 1801.

While it seems that the *Königlich-Preußische Akademie der Wissenschaften* [Royal Prussian Academy of Sciences] in Berlin would admit neither Bloch nor Mendelssohn because of their Jewish background, this may not have been an insurmountable blow, because Bloch had created his own club.[35] He was one of the founding members of the *Gesellschaft Naturforschender Freunde* [Society of Friends of Nature Research] in 1773.[36] As its name reveals, this society consisted of people united by their interest in studying nature. This club was, on the one hand, open and egalitarian, as it welcomed members from various backgrounds and religious denominations who wished to contemplate God's Creation through the study of nature. Furthermore, its statutes stressed that its members were considered equal without regard for birth, rank or standing.[37] On the other hand, the founders of the *Gesellschaft Naturforschender Freunde* were hesitant to admit people who were too affluent, as they feared that they would be too full of pride in their wealth and reputation to engage in debates in the spirit of equality and friendliness.[38] The membership of the *Gesellschaft* thus consisted largely of middle-class men, among them apothecaries, physicians, and government officials.[39] In order to join, these members had to demonstrate that they were serious *Naturkenner*.

A good way to do so was to collect natural historical objects. In fact, the ownership of a collection of natural rarities [*natürlichen Seltenheiten*] was a primary requirement for admission, as founding member Friedrich Wilhelm Heinrich Martini (1729–1778) declared.[40] In the spirit of collaboration and cooperation, the members' collections and libraries were to be made accessible to the others.[41] As Spary has argued for the Parisian collecting community, the *Gesellschaft Naturforschender Freunde* likewise preferred collections to be suitably ordered. As we saw, Nicolai's guidebook wrote approvingly of Bloch's *wohlgeordenete Naturaliensammlung*, which by and large adhered to Linnaean

35 Lesser, "Dr. Marcus Elieser Bloch," 242.

36 On the history of the *Gesellschaft Naturforschender Freunde*, see: Katrin Böhme, "Die Gesellschaft Naturforschender Freunde zu Berlin: Bestand und Wandel einer gelehrten Gesellschaft Ein Überblick," *Berichte zur Wissenschaftsgeschichte* 24, no. 4 (2001): 273.

37 Anke te Heesen, "Vom naturgeschichtlichen Investor zum Staatsdiener: Sammler und Sammlungen der Gesellschaft Naturforschender Freunde zu Berlin um 1800," in *Sammeln als Wissen. Das Sammeln und seine wissenschaftsgeschichtliche Bedeutung*, eds. Anke te Heesen and Emma Spary (Göttingen: Wallstein, 2001), 64.

38 Ludwik Lesser, *Chronik der Gesellschaft der Freunde in Berlin* (Berlin: Petsch, 1842), 46.

39 Katrin Böhme, *Gemeinschaftsunternehmen Naturforschung: Modifikation und Tradition in der Gesellschaft Naturforschender Freunde zu Berlin 1773–1906* (Stuttgart: Franz Steiner, 2005), 29.

40 Friedrich Wilhelm Heinrich Martini, "Gesetze der Hiesigen Gesellschaft," *Beschäftigungen der Berlinischer Gesellschaft Naturforschender Freunde* 1, no. 1 (1775): xxviii.

41 Nickelsen, *Draughtsmen, Botanists and Nature*, 109.

FIGURE 15 Portrait engraving of Marcus Élieser Bloch, Johann Conrad Krüger. Johann Georg Krünitz, *Oeconomische Encyclopädie* vol. 31 (Berlin: Joachim Pauli, 1778) UNIVERSITY LIBRARY LEIPZIG

systematics.[42] By ordering his cabinet in a systematic manner, Bloch – quite literally – showcased his knowledge of the Linnaean system. Bloch and collectors like him thus presented their expertise in the same breath as they presented their material possessions.

That Bloch was a significant figure in the learned landscape is underscored by the portrait engraving that the German physician and naturalist Johann Georg Krünitz (1728–1796) commissioned of him for the thirteenth part of his *Oeconomische Encyclopädie* (Figure 15), published in 1784.[43] As its title indicates, this series offered an encyclopaedic overview of all matters relating to 'oekonomie' – a term that encompassed all kinds of practical endeavour that contributed to socio-economic improvement in one way or the other.[44] Bloch deserved a place in this survey: as the text on the engraving declares, he was much loved as a physician, nationally renowned for his study of nature and an esteemed philanthropist.[45] Each of the portraits of eminent figures that Krünitz commissioned for his encyclopaedia convey something that is specific to the person depicted. In the case of Bloch, the scenic backdrop against which his portrait is placed signifies his interest in the study of nature. A pair of putti peruse natural historical volumes, one of them with a magnifying glass in hand; one can still make out the drawings of fish on the pages. They are a visual nod to the fish books that Bloch became widely known for.

When the last volume of his book series appeared in 1795, Bloch's fame extended far beyond the German states. He had earned memberships of learned societies such as the Royal Society in London and the *Muséum national d'Histoire naturelle* in Paris, as well as those of Leipzig, Göttingen, Utrecht and Frankfurt, amongst others.[46] And yet, Bloch accrued these international accolades despite leaving Germany just one time, in 1796, when – well into his seventies and a few years before his passing – he travelled to Amsterdam and Paris to attend auctions, visit collections and even meet the renowned naturalist

42 Nicolai, *Beschreibung*, 601.

43 The engraving was by the hand of the artist Johann Conrad Krüger (1733–1791), who also made engravings of other notable figures including Moses Mendelssohn and Krünitz himself. See: Johann Georg Krünitz, *Oeconomische Encyclopädie* (Berlin: Joachim Pauli, 1773–1796).

44 Phillips, *Acolytes of Nature*, 35.

45 "Als Arzt, beliebt; als Forscher der Natur, berühmt bey seiner Nation nicht nur / als Menschenfreund, wie Mendelssohn geschäzt, ist Bloch, dem Wahrheit dieses Denkmal setzt: durch ihren Freund, Krünitz."; it is not clear whether the term 'Nation' refers to the Jewish community, in its usage in the Hebrew Bible, or an envisioned German state. Bloch himself uses this term in the preface to his first volume to refer to the various German states.

46 *Allg. Nat. der Fische*, vol. 12 (Berlin: J. Morino, 1795), title page.

Georges Cuvier (1769–1832), certainly one of the most celebrated *savants* of his time.[47]

Towards the end of his life, therefore, Bloch had become an internationally renowned author and collector. His Jewish background had several consequences for his position in the social, cultural and intellectual echelons of Berlin: while he was part of a vibrant community of Jewish thinkers, he also faced forms of exclusion. It is possible that Bloch felt that he had something to prove, and that in assembling his collection he showed that he was fully able to participate – and perhaps better than most Christians – in the learned life of Berlin and beyond. It is also easy to imagine that he opted for a relatively underexplored area of natural history, one that carried less risk of creating rivalry with more established naturalists. In any case, through building up a collection of natural rarities he created a space for himself in the city's learned landscape, even co-founding an egalitarian society for which owning a cabinet of natural curiosities was a prerequisite. Those who visited his cabinet in the Spandauerstraße could admire Bloch's piscine treasures with their own eyes. By publishing the series of fish books, which was largely based upon drawings of the specimens in his collection, he circulated knowledge about his collection to the wider, learned community. This series of books will now be discussed.

2 Charting German and "Foreign" Fish

In the preface to the sixth volume of his natural history of fishes, published in 1787, Bloch announced that it would mark the conclusion of this series of books.[48] This did not mean he truly thought his project was finished. After all, he still had more than a hundred unpublished drawings of fish in his possession, executed in the most beautiful colours, and many of the specimens in his impressive collection in Berlin were yet to be depicted.[49] Lack of research material was not the issue here, nor did the book series fall short of attention. The problem was, rather, that many of its subscribers had not actually paid their dues. Bloch had funded the full publication process himself. He conceded that of the 20,000 Reichsthalers he had spent out of his own pocket – a staggering sum – he had earned only about half back.[50] In order to guard his family

47 Karrer, "Marcus Elieser Bloch," 137.

48 Marcus Élieser Bloch, *Allgemeine Naturgeschichte der Fische*, vol. 6 (Berlin: Realschule, 1787), sig. a2r; the series is henceforth abbreviated as *Allg. Nat. der Fische*.

49 *Allg. Nat. der Fische*, vol. 6, sig. a2r.

50 *Allg. Nat. der Fische*, vol. 8 (Berlin: J. Morino, 1791), sig. *2r.

from further impoverishment, he was forced to put his project on hold.[51] While this statement of financial duress may seem somewhat unusual to us, it is yet another example of how Bloch adhered to the cultural code of the learned community of his time. The idea was that one published books as a service to the society, and not even in the slightest hint of a pursuit of profit.[52] As we saw, publishing a well-made book was more of an investment in one's name. Upon altering his publication strategy, Bloch ultimately managed to proceed with the project and deliver the final six volumes, bringing the book series to actual completion with the twelfth.

What made these volumes particularly expensive to produce was the fact that each description was accompanied by a hand-coloured engraving. As we saw in previous chapters, Willughby and Ray's *Historia piscium* was a tricky enterprise for the Royal Society on account of its many engravings, whereas Linnaeus opted to include none in his own works to keep them affordable. Bloch clearly was not willing to stint on images. Even though he agreed with Artedi and Linnaeus on the crucial importance of classifying fish on the basis of physical marks, he did not think that their decontextualised manner of description was sufficient for the purpose of demarcating species. As he explained in the inaugural volume of his fish series, he had noticed that many of the fish that he came across, "could not be determined from the works of Linnaeus nor Artedi, nor of the older ichthyologists, because the descriptions in the first two are in respect to certain fish too short, and the latter are often unreliable because of their bad and unfaithful images."[53] The Jewish naturalist deemed an illustration essential to the proper representation of a fish.

Bloch's series of fish books followed the expansion of his collection in close step. The first volume appeared in 1782 under the title *Oeconomische Naturgeschichte der Fische Deutschlands*. Bloch paid for its publication out of his own pocket (or likely, that of his wife) and it was printed by the bookseller Hesse in Berlin. It contained thirty-seven species descriptions and an equal number of engraved, hand-coloured plates. The latter were published

51 At this time, Bloch had three children: a son (whose name remains unknown) from his marriage to Breinche Rintel (1747–1769) in 1765, a daughter named Rose from wedlock with Cheile Ephraim (*c.*1757–1780) whom he had married in 1774, and his daughter Rebecca after marrying Rahel Bendix (1767–1833) in 1784.

52 Phillips, *Acolytes of Nature*, 51.

53 "[...] dass sich viele Fische, [...] weder nach dem Linné und Artedi, noch nach den ältern Ichthyologen bestimmen liessen, da die Beschreibungen der erstern in Ansehung mancher Fische zu kurz, und leztre wegen der Verwechselungen der Namen und der schlechten und ungetreuen Zeichnungen, öfters unzuverlässig sind." *Allg. Nat. der Fische*, vol. 1, sig *2r.

in separate, bound volumes in a large folio format, so that all of the fishes' parts could be made clearly visible.[54] Two more volumes on German fish followed in 1783 and 1784; these were printed by Realschule Buchhandlung. When Bloch had collected and described the fish of the German states, he still had a lot of fish in his possession that as yet undescribed as they were not native to Germany. The descriptions and illustrations of foreign fish were published in nine parts as *Naturgeschichte der ausländischen Fische*. The first three of these appeared with the Realschule (1785–1787), the rest with the publishing house Johann Morino & Comp (1790–1795). The combined series of fish books became known as the *Allgemeine Naturgeschichte der Fische*. When the twelfth and final volume was published in 1795, the fish series had classified, described and depicted well over four hundred species.

Bloch's series of fish books is divided into 'German' and 'foreign' fishes. This polarity mirrors the tendency of naturalists in the sixteenth and seventeenth centuries to contrast the 'indigenous' with the 'exotic'.[55] For botanists, however, the fruitful transportation of plants from the tropics to Europe was a topic of both considerable scientific and economic interest.[56] Acclimatisation was difficult, but the successful cultivation of plants like the Peruvian potato and American tobacco in the cold Dutch Republic indicated that boundaries between 'indigenous' and 'exotic' were permeable.[57] As it turns out, Bloch also saw great opportunities for introducing fish from the warmer to the cooler climes. His book contained detailed advice on how one should go about transporting live fish from one continent to the other, recommending the selection of fish that were close to spawning and to emulate the original environment of the fish as closely as possible in their new home. With the exception, however, of hardy species like the goldfish, a beloved creature from East Asia that Bloch noted had been successfully naturalised in Amsterdam, London and Berlin, such attempts to transplant fish rarely proved successful.

The contrast between Bloch's proximity to the German fish he described in his work on the one hand, and his distance from the foreign fish that he included on the other has caught the attention of historians. As Johannes Müller has argued, Bloch's examination of German fishes and of foreign fishes followed a different methodology and approach. For the German fish, he

54 *Allg. Nat. der Fische*, vol. 1, sig. *3r.

55 Cooper, *Inventing the Indigenous*, 32.

56 See: Staffan Müller-Wille, "Walnuts at Hudson Bay, Coral Reefs in Gotland: The Colonialism of Linnean Botany," in Schiebinger and Swan, *Colonial Botany*, 34–48.

57 Alix Cooper, "The Indigenous versus the Exotic: Debating Natural Origins in Early Modern Europe," in *Landscape Research* 28 (2003): 58.

would often have the opportunity to examine various living specimens with his own eyes. In his work on the foreign fishes, Bloch had to interpret the observations of others, evaluating and comparing the different written accounts at his disposal before settling upon his own verdict.[58] What complicated matters was that Bloch's knowledge of geography came from books rather than from travel, and that at times he attributed species to their wrong locales, confusing, for example, species from South America with those from South India.[59] His distance from the environments of these foreign fish posed another epistemological challenge. As Dorothee Fischer has shown, it was difficult for naturalists such as Bloch to reconstruct how a species would have looked like while still alive based on prepared specimens.[60] Where the last section of this chapter looks more closely at matters of interpretation, this section examines how Bloch envisioned the contribution of his own work and how he sought to cater to different audiences.

When the Holy Roman Emperor Joseph II (1741–1790) received the first volume of Bloch's book series, he awarded him with a gold medal in recognition of what he perceived to be the book's great benefits.[61] Bloch presented the three first volumes of his series of fish books as an oeconomic natural history. In his preface, Bloch noted, not without astonishment, that while entire societies had dedicated themselves to mastering the intricacies of bee-keeping, fish received only scant attention. "Do fishes not equally deserve", Bloch wondered, "our attention; do they not form an important part of our diet; have they not always been an important trade stuff?"[62] The German fish were certainly deserving of a series of their own. Bloch wanted his work to be useful not only to scholars [*Gelehrte*], but also to agriculturists [*Landwirthe*], and he therefore included a discussion of different types of fishing nets and how to use them.[63] Bloch indicated which net to use for each fish, explained at what time of the year it

58 Johannes Müller, "Distance, Geography, and Anecdote in M.E. Bloch's Natural History of Fishes," in Smith and Egmond, *Ichthyology in Context*, 628.

59 Ibid.

60 Dorothee Fischer, "The Afterlives of Fish Far from Home: (Mis)Representations in the Iconography of Preserved and Printed Pufferfish in 18th-Century Germany," in Smith and Egmond, *Ichthyology in Context*, 566.

61 Notice in *Magazin des Buch- und Kunsthandels, welches zum Besten der Wissenschaften und Künste von den dahin gehörigen Neuigkeiten Nachtricht giebt* (Leipzig: Johann Gottlob Immanuel Breitkopf, 1782) no. 7, 558.

62 "[…] verdienen aber die Fische nicht eben so wohl unsre Aufmerksamkeit; machen sie nicht einen grossen Theil unsrer Nahrung aus; waren sie nicht zu allen Zeiten ein wichtiger Handlungszweig?" *Allg. Nat. der Fische*, vol. 1, sig. *2v/*3r.

63 Ibid., 1.

was best to catch it and how it should be prepared; information he might have gathered during the summer vacations that he spent in a nearby fishing village in order to collect useful observations from the fishermen.[64] Rather than catering to a strictly Latinate audience, as Willughby, Ray and Artedi had done in their works of fish, Bloch decided to publish his fish series first and foremost in the vernacular, thus making it accessible to a wider, German-speaking public. Again, this was very much in line with the broader ideal shared by communities of *Naturforscher*, namely that one produced a work as an act of service to the community.[65]

In the first volume of the series, Bloch gave his own definition of fish: "I take the word *fish* in its common parlance, and understand by it all those water dwellers that move through their element with fins. To it [fish] therefore belong also the whales and swimming amphibians, which Linnaeus saw fit to separate from the fish in the twelfth edition of his natural system."[66] Interestingly, however, Bloch did not actually discuss the whales in his books because he was aware that his fellow naturalist Johann Christian von Schreber (1739–1810) was already planning on incorporating them in his natural historical series on mammals.[67] This goes to show how the meaning of the word 'fish' continued to fluctuate even after Linnaeus had presented his ultimate system of the order of nature, with some naturalists placing whales in the realm of fishes and others in the realm of mammals.

Just as the fishes on the shelves of Bloch's cabinet were ordered according to Linnaean principles, so were the fishes described in his book series. He conceived his work as a continuation of the taxonomical system that Linnaeus had presented in the twelfth edition of his *Systema naturae*.[68] The classification system for fish in this edition diverged from that of the first edition, which had effectively presented the system of Artedi, in a few respects. Most changes had occurred on the level of orders, all of which Linnaeus replaced with his

---

64 Ibid., sig *3r.

65 Phillips, *Acolytes of Nature*, 37.

66 "Ich nehme das Wort *Fisch* nach dem gewöhnlichen Sprachgebrauche und verstehe darunter alle diejenigen Wasserbewohner, welche sich mittelst der Flossen in ihren Elemente bewegen. Es gehören daher auch die Wallfische und schwimmende Amphibien mit in meinen Plan, welche Linné in der zwölften Ausgabe seines Natursystems davon zu trennen für gut fand." *Allg. Nat. der Fische*, vol. 1, 2. Contrary to Bloch's claim, however, this separation had been suggested already in the 10th edition of the *Systema naturae*.

67 Johann Christian von Schreber, *Die Säugthiere in Abbildungen nach der Natur mit Beschreibungen* (Erlangen: Wolfgang Walther, 1774–1804).

68 Carl Linnaeus, *Systema naturae*, ed. 12 (Stockholm: Lars Salvi, 1766–1767).

own.[69] He also added new genera.[70] The underlying aims and principles were not altogether different from those of the *Ichthyologia* – it was just that other characteristics were selected as the basis for certain taxonomical ranks. The number of fins and their relative position to one another, as well as the number of rays in a fin, remained especially salient features for deciding genera and species. By the last volume of the *Allgemeine Naturgeschichte der Fische*, Bloch had expanded the list of known fish by no fewer than 250 species.[71]

Bloch's descriptions adhered to the same general format. They were headed by the vernacular name of the fish, followed by its Linnaean binomial, a reference to its corresponding plate, and one sentence describing its main characteristics. These included one sentence enumerating the number of rays in each and every fin in an abbreviated and almost formulaic manner: in the case of a species of flying fish, for example, he noted 'K. 10, Br. 15, B. 6, A. 13, S. 20, R. 17', referring to the number of bones in the gill flap and the number of rays in the pectoral fin, pelvic fin, anal fin, caudal fin and dorsal fin respectively.[72] It is an example of the quantitative focus that had governed Artedi's work, and had also been visible in that of Willughby and Ray. If the fish had already been described by earlier authors, Bloch listed all of these, from Aristotle to Artedi – demonstrating his comprehensive knowledge of natural historical literature and, by extension, offering a glimpse into his well-stocked library.

The first paragraph of the actual species description gave a more elaborate description of the fish's main characteristics. This focus on differentiating marks was also seen in the works Willughby and Ray, Artedi and Linnaeus. Each species description included discussions of the external parts of the fish; if Bloch was able to perform a dissection of the species, these descriptions were supplemented by reports on their inner parts. In some instances, the descriptions contained statements on how the fish at hand could best be caught and with which tools, when it was best eaten, and even, in a few cases, recipes. In line with Bloch's preoccupation with the oeconomic qualities of fish, the procreation and growth of fish were also important topics. Bloch's

---

69 Linnaeus had relegated the order of the *Plagiuri* to his class of mammals, and that of the *Chondropterygii* to the amphibians. He did away with the *Malacopterygii* and *Acanthopterygii* (a division based on soft or thorny rays). His newly established orders *Apodes, Jugulares, Thoracici,* and *Abdominales* were all based on the presence and position of pelvic fins. Linnaeus, *Systema naturae*, ed. 12, 422.

70 See: *ibid.*, 423–424.

71 According to an estimation cited in Wells, "M.E. Bloch's Allgemeine Naturgeschichte der Fische: A Study," 9.

72 Currently accepted species name: *Exocoetus evolans. Allg. Nat. der Fische*, vol. 12, 14. The German names for these fins are *Kiemenhaut, Brustflosse, Bauchflosse, Schwanzflosse, Rückenflosse.*

descriptions varied in length. For instance, the species description of a carp that he had received from Malabar barely took up one page, whereas that of the common carp, a fish well-known across the German states, stretched out over fifteen pages and even contained an elaborate account of how they are bred in ponds.[73] All species descriptions concluded with a summary of the names given to the species in other languages.

On describing a species, Bloch first ascertained whether earlier naturalists had already described it. If that was the case, he would correct the existing accounts where necessary. For example, he stated how the caudal fin of the carp consisted of 11 rays rather than the 9 rays that Artedi had reported.[74] Bloch speculated that Artedi must have copied this number from Willughby, who neglected to include the smaller rays at the end of the tail in his count. In seeking to refine the existing classification system as well as he could, Bloch did not, however, follow Linnaeus and Artedi to the letter. He accorded weight to characteristics that the Swedish naturalists had not cared for. They, for example, had only occasionally commented on the colouration of fish in their species descriptions, as they deemed colour too unstable a quality for use in classification. Artists and naturalists alike grappled with the subjective qualities of colour, and attempts to codify it either visually or textually proved complicated at best.[75] Nevertheless, Bloch considered colour to be a valuable characteristic for recognising species and paid it a lot of attention in both word and image. In some instances, he took colouration as the defining mark of the fish, such as in the case of a new species he described and which he named the red mackerel.[76] "The red colour", Bloch wrote, "sets this fish apart from the others of this department [the genus]."[77] He continued "[o]n the back and on the sides until the lateral line, the red colour predominates, through which the silver colour shines through, but from thereon however, the proportion is the other way around. The fins are yellow, and play into purple."[78] As will be dis-

---

73 The Latin name Bloch gave this fish was *Scomber ruber*; its currently accepted species name is *Cyprinus carpio*.

74 *Allg. Nat. der Fische*, vol. 1, 44.

75 On these efforts, see, for example, Richard Mulholland, "The Mechanism and Materials of Painting Colour *ad vivum* in the Eighteenth Century," in Balfe, Woodall and Zittel, *Ad vivum?*, 328–3355; and Joachim Rees, *Die Verzeichnete Fremde. Formen und Funktionen des Zeichnens im Kontext europäischer Forschungsreisen 1770–1830* (Paderborn: Wilhelm Fink, 2015), esp. chapter 4, "Kodiertes Kolorit," 153–230.

76 Currently accepted species name: *Caranx ruber*.

77 "Die rothe Farbe unterscheidet diesen Fisch von den übrigen dieser Abtheilung." *Allg. Nat. der Fische*, vol. 10, 75–76.

78 "Am Rücken und an den Seiten bis an die Seitenlinie hat die rothe Farbe die Oberhand, durch welche die Silberfarbe durchschimmert, von da weiter aber, verhält es sich umgekehrt. Die Flossen sind gelb, und spielen in's Violette." Ibid.

cussed in the last part of this chapter, he exhorted both himself and his artists to bring out colours the best way possible in the illustrations.

Bloch may have been intent on making his books of interest to a wide audience, but it seems like not everyone would have been able to afford it. An indication of how much one might have had to pay for a copy can be found in the price listings of late eighteenth-century sales catalogues. In 1792, for example, the publisher Morino put out an advertisement for the ninth volume of the fish series, which he announced would cost the same as the preceding volumes: 10 Reichsthaler for the folio, or 12 Reichsthaler for a large folio.[79] That same year, a bookseller in Jena offered the 9 volumes that had been published so far as a set at the discount rate of 70 Reichsthalers rather than the 120 he claimed it usually went for.[80] When Leipzig bookseller Johann Gottlob Beygangs (1755–1823) advertised the book series in 1797, he asked between 12 and 18 Reichsthalers per volume, depending on both the size and quality of the paper it was printed on.[81] These examples give a sense of the price range of the series – buyers should expect to pay at least 10 Reichsthalers per volume unless a seller made a sensational offer. For a fisherman or farmer, who Bloch had mentioned as possible readers, such a price would have been steep if not altogether out of reach. In comparison: the aforementioned Beygangs priced an exegetical handbook of the Old Testament at about two thirds of a Reichsthaler.[82] The capital investment represented by Bloch's books of fish also demanded no little caution on the part of the bookseller. The bookseller Beygangs concluded his advertisement for the series of fish books with a nota bene: "[…] this important work can, due to its considerable cost, henceforth no longer be bought on credit, but must be paid for in cash."[83]

From the eighth volume (1791) onwards, plates are inscribed with the names of those who financed the engraving. Bloch opted for this publication strategy when production costs became too onerous, citing Willughby and Ray's *Historia piscium* as an exemplar.[84] The archive of the *Berlin-Brandenburgische*

79 Advertisement by Johann Morino, *Intelligenzblatt der Allgemeinen Literatur-Zeitung vom Jahre 1792* (Leipzig: Johann Gottfried Müllerischen Büchhandler, 1792), 338–339.

80 Advertisement by Hn. Adv. Fiedler, *Intelligenzblatt der Allgemeinen Literatur-Zeitung vom Jahre 1792*, 767.

81 Advertisement by Johann Gottlob Beygangs, *Intelligenzblatt der Allgemeinen Literatur Zeitung vom Jahre 1797* (Leipzig: Johann Gottfried Müllerischen Büchhandler, 1797), 246.

82 Ibid., 245.

83 "NB. Dieses wichtige Werk kann wegen des zu grossen Kostenaufwand, fernerhin nicht mehr in Rechnung, sondern gegen gleich baare Bezahlung erlassen werden." Advertisement by Johann Gottlob Beygangs, *Intelligenzblatt der Allgemeinen Literatur-Zeitung vom Jahre 1792*, 348.

84 *Allg. Nat. der Fische*, vol. 8, sig. *3r.

Akademie der Wissenschaften holds a circular encouraging readers to take out plate subscriptions for Bloch's series.[85] This undated leaflet was probably a response to Bloch's 1787 announcement that he would cease publication. The document, likely drafted by scholars and publishers close to Bloch, is addressed to friends and patrons of learning. It suggested that "every supporter of this enterprise underwrite the cost of producing as many plates, at 2 Louis d'Or apiece [around 10 Reichstalers], as his desire to see this work completed inspires him."[86] The circular offers a fascinating insight into the reasons why the series was considered worthwhile. It explains that Bloch's fish series, "a work whose accuracy has earned it the best reviews in Germany, France, England and every foreign country, and has been universally been declared a classic work of natural history", might otherwise be left uncompleted.[87]

This, the authors declared, would be a shame. It could take centuries, they stated, before another scholar as well positioned as Bloch to combine a unique possession of materials with the right approach to the subject matter might emerge.[88] Subscriptions were solicited "for only 200 plates, [...] with which this work will be completed for Germany's honour."[89] Judging from the number of plates on which names are engraved, the appeal was successful. The *Königlich-Preußische Akademie der Wissenschaften*, for example, although unwilling to admit Bloch as a member, was prepared to support his work financially.[90] As can be seen on (Figure 14), the Jewish banker Bernhardt Friedländer, brother of the aforementioned David, also sponsored a plate.

85 Circular in Berlin-Brandenburgische Akademie der Wissenschaften (hereafter BBAW), Berlin, PAW (1700–1811), I-XII-11.

86 "[...] dass jeder Beförderer dieses Unternehmens die Kosten für so viele Platten, eine jede zu 2 Louisd'or zu bezahlen unterchreibt, als ihm seine Liebe zu der Vollendung dies Werks eingiebt." BBAW, PAW (1700–1811), I-XII-11, f1v. This equivalent is given in the newspaper Der Anzeiger: Ein Tagblatt zum Behuf der Justiz, der Polizey und aller bürgerlicher Gewerbe, wie auch zur freyen gegenseitigen Unterhaltung der Leser über gemeinnützige Gegenstände aller Art 19, no. 19–20 (1792): 154.

87 "[...] ein Werk, das sowohl, wegen seiner Richtigkeit, in Deutschland als auch in Frankreich, England und jedem Auslande die besten Recensionen erhalten hat, und durchgängig in der Naturgeschichte für ein klassisches erklärt wird [...]" ibid., f1r.

88 "[...] und Jahrhunderte verstchreichen, ehe wieder ein Gelehrter aufstände, wo sich Besitz der Materialien und der nehmliche richtige Standpunkt auch so vereinigten, wie bey dem in diesem Fache schon so rühmlich bekannten Herrn Doktor Bloch?" Ibid.

89 "Für 200 Platten wird nur Subscription angenommen, als womit dieses Werk zu Deutschlands Ehre vollendet seyn wird." BBAW, PAW (1700–1811), I-XII-11, f1v.

90 Plates CCCXXVIII–CCCXXX, CCCXXXII–CCCXXXIII, CCCXXXV, CCCXXXVIII–IX, Hans-Joachim Paepke, "Ein jüdischer Untertan des Preußenkönings Friedrich II. studiert die Fischfauna der Welt," in *Klasse, Ordnung, Art: 200 Jahre Museum für Naturkunde*, eds. Ferdinand Damaschun, Sabine Hackethal, Hannelore Landsberg, and Reinhold Leinfelder (Rangsdorf: Basiliskenpresse, 2010), 87.

That Bloch wished to reach as wide of an audience as possible is evident from his plans for translating his book series. He arranged the translation of the series into French not long after the first volume had appeared in German.[91] These translations appeared between 1785 and 1797 as *Ichtyologie, ou, Histoire naturelle, générale et particulière des poissons*, and were published by François de la Garde in Berlin.[92] The French edition follows the original German very closely: it retains its focus on German fish and does not contain a separate preface introducing the translation.[93] French was, of course, a language more widely read in international learned circles than Bloch's German, and thus made his work accessible to a larger scholarly public. In 1787, Bloch's son journeyed to France and England to seek subscribers for the French edition.[94] In June 1791, Bloch wrote to Joseph Banks to solicit his opinion on also issuing an English translation. In his reply, Banks explained that while he was charmed by the idea "par amour de la science Ichtiologique" [*sic*], the British "Gens de lettres" for the most part understood either French or German or both so that it was doubtful whether he could find the number of subscribers needed for such a translation.[95] After the Jewish naturalist's death, a Dutch translation was begun but never completed.[96]

As this section has shown, Bloch's series of fish books spanned various territories and audiences, no matter that they sometimes appeared at odds. His initial aim was to chart all the fishes of the German states and, offering descriptions rich with morphological detail in the vernacular that they might benefit fishermen and farmers as well as scholars. He included illustrations to facilitate recognition of species, more so than the diagnostic sentences that Linnaeus and Artedi provided. Bloch started to build up a picture of this region by collecting its fish. He then expanded his scope to include "foreign" fishes

91 The translation was carried out by Jean Charles Thibault de Laveaux (1749–1827), professor of French in Basel. Ellen B. Wells, "M.E. Bloch's Allgemeine Naturgeschichte der Fische: A Study," *Archives of Natural History* 10, no. 1 (1981): 7–8.

92 Marcus Élieser Bloch, *Ichtyologie, ou, Histoire naturelle, générale et particulière des poissons*, trans. Jean Charles Thibault de Laveaux (Berlin: François de la Garde, 1785–1797).

93 The main difference appears to be that, in the German edition, descriptions and engravings were bound separately, while in the French edition, the engravings interleave the descriptions.

94 Karrer, "Marcus Elieser Bloch," 136.

95 Joseph Banks to Bloch, dated 24 June 1791, Abteiling Historische Drucke of Staatsbibliothek zu Berlin (hereafter SBB), Berlin, Sammlung Darmstaedter Weltreisen 1768: Banks, Sir Joseph, fır.

96 The first part was printed by Cornelis Nozeman and Johann Christiaan Sepp under the title *Afbeeldingen en beschrijvingen van in- en uitlandsche visschen M.E. Bloch; gevolgd naar het Hoogduitsch* in Zaltbommel in 1804. I thank Esther van Gelder for drawing my attention to it.

and issued a French translation of the entire series. He sought out opportunities to further distribute his work, eventually choosing the early modern equivalent of crowd-funding, subscription, when the cost of producing its hand-coloured engravings became too much for him to bear alone. Bloch's series of fish books connected several geographical territories, both national and global. One important geographical context was colonial, and it is to this that we now turn.

3 Colonial Collecting on the Coast of Coromandel

The many ways in which the global entanglements of empire and trade spurred maritime knowledge-making have been evident in previous chapters. The dried sea-unicorn from Brazil in Grew's cabinet, Sloane's account of his voyage to Jamaica, the siren allegedly caught and transported to Leiden by a Dutch WIC official, Seba's collection of natural treasures from the Dutch East Indies, or Catesby's book of colourful fishes from Carolina, Florida and the Bahamas all bear its fingerprints. Bloch's oeuvre of fish provides a particularly interesting case study of how knowledge of nature was made in the colonies. Given that the early modern German states did not have a continuous presence outside Europe, the colonial connections of his work is less overt than in some of the abovementioned natural historical publications. And yet, as we will see, a significant part of Bloch's collection was created in missionary settlements overseas, with local informants and assistants involved in every step of the way – from collecting, preserving, drawing, to describing and packing. Both his collection and the series of volumes derived from it serve to illustrate that it was not merely commercial aims that worked to expand global infrastructures of knowledge, but that religious aspirations were anything but a negligible factor in expanding global infrastructures of knowledge.[97]

If we look at the origins of the 'foreign fish' that were described in Bloch's book series, one particular source stands out. The German Pietist missionary Christoph Samuel John, introduced in the beginning of this chapter, was Bloch's top supplier in this category. John is mentioned as a donor in over fifty species descriptions, which amounts to more than a tenth of the total

97 See, for example, Andrés I. Prieto, *Missionary Scientists: Jesuit Science in Spanish South America, 1570–1810* (Nashville: Vanderbilt University Press, 2011); Florence C. Hsia, *Sojourners in a Strange Land: Jesuits and Their Scientific Missions in Late Imperial China* (Chicago: University of Chicago Press, 2009).

of species discussed in the series.[98] John was stationed in Malabar in South India, on the Coast of Coromandel. He had reached out with an offer to help collect specimens after acquiring one of Bloch's books from the *Nachlass* of his late fellow missionary (and former student of Linnaeus) Johann Gerhard König (1728–1785).[99] John's correspondence in the archives of the *Franckesche Stiftungen* [Francke Foundations] and the published reports from his hand show how assisting Bloch served his missionary purpose.[100] After studying Theology at the University of Halle, John worked at the Foundations erected by the Lutheran pastor August Hermann Francke (1663–1727) in Halle an der Saale 1698 in order to educate and elevate orphans through Pietist faith.[101] Francke's religious and societal ideals stretched far beyond the city's borders. He had established the Danish-Halle mission at the Coast of Coromandel in South India in 1702, at the behest of the Danish King Frederick IV (1671–1730). In 1771, John boarded ship to become part of this mission, settling in a Danish colony named Tranquebar, a fishing village that was called Tarangambâdi, or 'village of the singing waves' in the Tamil language.[102]

When John landed at Coromandel, he found a land that had been shaped by imperialism. The Portuguese had made settlements on the Coast of Coromandel in the sixteenth century, and they were soon followed by Dutch traders. The British East India Company also established a trading port here in the late seventeenth century. Such settlements often rested on the labour of enslaved Africans or Southeast Asians, transported there to be forced to work

98 Arthur MacGregor, "European Enlightenment in India: An Episode of Anglo-German Collaboration in the Natural Sciences on the Coromandel Coast, Late 1700s–Early 1800s," in MacGregor, *Naturalists in the Field*, 378.

99 Christoph Samuel John to Johann Ludwig Schulze, 18 October 1787, AFSt/M 1 C 28 : 87; Anne-Charlott Trepp, "Matters of Belief and Belief that Matters: German Physico-Theology, Protestantism, and the Materialized Word of God in Nature," in Blair and Von Greyerz, *Physico-Theology*, 135.

100 John's correspondence with his superiors is held by the *Frankesche Stiftungen* in Halle an der Saale as part of their Missionsarchiv mit der Indien- und der Amerikaabteilung, AFSt/M. His reports frequently appeared in the *Neue Hallesche Berichten*, the printed periodical of the mission. A general discussion of source material regarding the mission can be found in Erika Pabst and Thomas Müller-Bahlke, *Quellenbestände der Indienmission 1700–1918 in Archiven des deutschsprachigen Raums*.

101 For a history of the *Frankesche Stiftungen*, see: Holger Zaunstöck, ed., *Gebaute Utopien, Franckes Schulstadt in der Geschichte europaischer Stadtentwürfe* (Halle: Frankesche Stiftungen, 2010).

102 Daniel Jeyaray, "Mission Reports from South India and Their Impact on the Western Mind: The Tranquebar Mission of the Eighteenth Century," in *Converting Colonialism: Visions and Realities in Mission History, 1706–1914*, ed. Dana L. Robert (Cambridge: William B. Eerdmans, 2008), 23.

in planting, farming and fortification.[103] In the area where John was stationed, he would have encountered these unfree labourers, as well as Tamil of different castes and professions or traders, officers, and missionaries of different European nationalities. It was against this background that John took up his work at the mission: converting local inhabitants and teaching at the mission schools which were attended by both European and Indian pupils.

Rather than reciting or discussing Scripture, John came to consider the study of nature as an effective tool of conversion.[104] He began, for example, to teach the school children botany.[105] This would not have been an altogether new idea for John: the Halle orphanage where his teaching had commenced was furnished with a *Kunst- und Naturalienkammer* through which its pupils could contemplate the wisdom of God.[106] Slowly but surely, John amassed a collection of natural specimens containing aquatic animals, reptiles, amphibians and insects as well as stuffed birds and mammals.[107] The collecting of *naturalia* was a widespread practice among the various European mission posts in India such as the United Brethren of the Moravian church.[108] Besides König and John, the missionary Johann Peter Rottler (1749–1836), for example, was active in trading specimens. These and other nature-minded missionaries met each other as well as government officials and private merchants in the Tranquebarian Society, which was founded in 1788.[109]

The exclusive access to the natural riches of Tranquebar and surrounding islands these missionaries enjoyed made them invaluable for European naturalists. There was an intricate infrastructure in place through which several collectors on different continents could exchange objects. John not only collected

---

103 Anna Winterbottom, *Hybrid Knowledge in the Early East India Company World* (Basingstoke: Palgrave Macmillan, 2015), 169, 183.

104 Karsten Hommel, "Physico-Theology as Mission Strategy: Missionary Christoph Samuel John's (1746–1813) Understanding of Nature," in *Halle and the Beginning of Protestant Christianity in India*, vol. 3, eds. Andreas Gross, Y. Vincent Kumaradoss, and Heike Liebau (Halle: Franckesche Stiftungen, 2006), 1115.

105 Heike Liebau, *Cultural Encounters in India: The Local Co-Workers of Tranquebar Mission, 18th to 19th Centuries* (London: Routledge, 2017), 399.

106 See: Stefan Laube, "Privilegierte Dinge für Unterprivilegierte? Die Kunstkammer im Waisenhaus," in Dolezel, Godel, Peča and Zaunstöck, *Ordnen – Vernetzen – Vermitteln*, 49–72.

107 Hommel, "Physico-Theology," 1112.

108 John was inspired by the collection of the Herrnhut missionaries, which also functioned as a deposit of objects for his own collection. Thomas Ruhland, *Pietistische Konkurrenz und Naturgeschichte: Die Südasienmission der Herrnhuter Brüdergemine und die Dänisch-English-Hallesche Mission (1755–1802)* (Herrnhut: Herrnhuter Verlag, 2018), 256.

109 Niklas Thode Jensen, "The Tranquebarian Society: Science, Enlightenment and Useful Knowledge in the Danish-Norwegian East Indies, 1768–1813," *Scandinavian Journal of History* 40, no. 4 (2015): 535.

for Bloch, but also sent items to other collectors including the Danish preacher and naturalist Johann Hieronymus Chemnitz (1730–1800), who also seems to have acted as John's agent in transporting specimens throughout Europe.[110] Some of these missionaries divided the realm of nature among them. In a letter to his superior in Halle, Johann Ludwig Schulze (1734–1799), John revealed that he agreed with Rottler that the latter would concern himself with the field of botany, and that he would focus on the study of animals.[111] As he noticed that hitherto fish had not been deemed a subject particularly deserving of attention, making it relatively easy to stumble on as yet undescribed species, he decided they would be his focus.[112] As John wrote to his superior: "I was most eager to see how far the East Indian fish were known, and to support his [i.e., Bloch's] work through bringing to light more [of them]."[113]

But who did the actual collecting of these specimens? Colonial contributors to European collections often drew on the know-how and expertise of local collectors.[114] Delbourgo has shown, for example, that Sloane relied on local assistants, including African and (West) Indian hunters, fishermen, and divers to obtain Jamaican animals and that he wrote appreciatively of their methods, such as the manner in which Africans caught fish by intoxicating them with dogwood-bark.[115] John likewise depended on local fishing communities to furnish him with specimens. In a letter to Bloch in February 1792, published in the enlightened monthly magazine *Berlinische Monatschrift*, he describes organising a fish-collecting session in which fishermen took specimens from the mission garden ponds, the rivers, and the sea.[116] In other instances, his European and Tamil pupils collected fishes.[117] John furthermore possessed a considerable number of shells from the port city of Tuticorin (now Thoothukudi) on the Bay of Bengal, some of which had been procured for him by Tamil divers.[118] When it came to collecting the world underwater, diving was a particularly prized skill for Europeans, few of whom knew how to swim. Delbourgo has described how in Isla Margarita near Venezuela, first indigenous American divers and

110 On Chemnitz, see: Trepp, "Matters of Belief and Belief that Matters," 132.

111 John to Schulze, 20 January 1790, AFSt/M 1 C 31a : 21.

112 Ibid.

113 "Ich war just am begierigsten darauf um zu sehen wie weit die Ostindische Fische bekannt wären und sein Werk durch die Bekantmachung mehrere zu unterstutzen." John to Schulze, 12 October 1789, AFSt/M 1 C 30a : 2.

114 Barlow Robles, *Curious Species*, 150.

115 Delbourgo, *Collecting the World*, 118.

116 Christoph Samuel John, "Einige Nachrichten von Trankenbar auf der Küste Koromandel. Aus einem Briefe von dem Missionarius Hrn John an Hrn Doktor Bloch in Berlin," *Berlinische Monatschrift* 20 (1792): 587.

117 John to Schulze, 20 January 1790, AFSt/M 1 C 31a : 21.

118 MacGregor, "European Enlightenment in India," 377.

later African divers were forced to retrieve maritime treasures such as pearls from the bottom of the sea, and how their physical integrity was compromised due to too long and frequent dives.[119]

Maritime knowledge-making in the colonies might have entailed such unfree work, but its exact nature is not always easy to reconstruct. In tracing "the labour behind Linnaeus's monumental storehouse", Barlow Robles provides the example of a black servant who collected several of the fish that Alexander Garden would later send to Linnaeus, explaining that this unnamed individual was likely enslaved.[120] Language can obscure relations and identities. Anna Winterbottom has noted this for the records of the EIC, where the words 'black servants' and 'slaves' are used interchangeably, and the term 'black' is used in reference both to Africans and Asians.[121] We also encounter terms such as 'servants' and 'black' in the letters of John. He referred, for example, to those helping him collect and prepare specimens as 'black hands'.[122] John appointed one of his local assistants, a 'black Natural Philosopher', on an expedition to the Nicobar Islands to collect fish and shells, as well as insects, seeds and plants.[123] Although the designation of natural philosopher conveys status to this individual, much is left unsaid about his background and the nature of this appointment. While John's mission activities are relatively well documented, we often lack proper insight regarding what was at stake for those individuals who assisted him. It is not always clear what they would stood to gain by this and whether their involvement in these activities were a matter of choice, coercion, or something in between.

Once a specimen had been collected, John gathered information that he felt Bloch might use in his descriptions. He would often include local names of the species, as well as elaborations of how they were caught, how they tasted and how they could be prepared and preserved as a foodstuff. John's descriptions range from brief notes to rather more extensive reports which offer a glimpse of fishing activities in the region. A species of the order of blennies, for example (called *Karumudel* in Tamil) that dwelled in the sea near river mouths and was often caught in November with fishing rods, measured around fifteen inches in length and had fat and tasty flesh. Its meat was consumed both fried and boiled, and it was sometimes dried to last longer.[124] On the species

119 Delbourgo, "Divers Things," esp. 159–176.

120 Barlow Robles, *Curious Species*, 150.

121 Winterbottom, *Hybrid Knowledge*, 165.

122 Christoph Samuel John, 'Einige Nachrichten von Trankenbar' in: *Berlinische Monatsschrift* 22 (1794) 350–351.

123 Pratik Chakrabarti, *Materials and Medicine Trade, Conquest and Therapeutics in The Eighteenth Century* (Manchester: Manchester University Press, 2023), 120.

124 *Allg. Nat. der Fische*, vol. 12, 101.

called *Kalamin*, the Tamil name for the common threadfin, John wrote in the description accompanying a specimen that it could grow up to four foot in length, weighing so much that one required strong muscles to carry it.[125] The fish was dried and salted, or sometimes cooked and eaten with wine vinegar.[126] Such knowledge, which Bloch was so keenly interested in for his German fish, stemmed not only from John's own observations, but also came from local fishermen and cooks skilled in preparing and preserving these fish. John also carefully noted the taxonomic divisions that the Tamil made between four species of pomfret, of which Bloch had so far only described one.[127]

When possible, John had drawings made of the collected specimens, and even mentions enlisting an Indian *Zeichenmeister* for this purpose; unfortunately, none of the drawings that John sent to Bloch seem to be extant.[128] He also instructed his European pupils to draw the collected *naturalia*, but he found they often did not know how to draw 'after nature', and that they would invariably leave once they had mastered this skill to some degree, much to his dismay.[129] With suitable drawing skills in short supply, those possessing this ability could put this asset to market, as John describes:

> It is unfortunate, however, that my draughtsman has left me, and has gone into English service, where he is paid around 80 gold reichsthaler monthly. I did all I could to keep him with me, and offered him 1 reichsthaler for every drawing. He laughed at me, however, for such a small thing; even though he is only the son of a sergeant, has a black mother, and himself looks like burnt coffee.[130]

---

125 Currently accepted species name: *Polydactylus plebeius*.

126 Allg. Nat. der Fische, vol. 12, 24.

127 Ibid., 89.

128 Karrer, "Marcus Elieser Bloch," 146.

129 Christoph Samuel John, "Fragen des Herrn Professor Forster in Halle an die Missionarien in Trankenbar, und Herrn Johns Antworten darauf," *Neuere Geschichte der evangelischen Missionsanstalten zu Bekehrung der Heiden in Ost-Indien* 4, no. 43 (1793): 655; John, "Einige Nachrichten von der Küste Koromandel," 352.

130 "Traurig aber ist es, dass mein Zeichner mich verlassen hat, und in Englische Dienste gegangen ist, wo er monatlich gegen 80 Rthlr. Gold erhalt. Ich that alles Mögliche, ihn bei mir zu behalten, und bot ihm für jede Zeichnung bis 1 Rthlr. Er lachte mich aber für einde solche Kleinigkeit gleichsam aus; ob er gleich nur eines Unteroffiziers Sohn war, eine schwarze Mutter hatte, und selbst wie gebrannter Kaffee aussah." John, "Einige Nachrichten von Trankenbar auf der Küste Koromandel. Aus einem Briefe von dem Missionarius Hrn John an Hrn Doktor Bloch in Berlin," *Berlinische Monatschrift* 20 (1792): 589–590.

This remark makes clear just how prevalent hierarchical ideas with regard to skin colour were, with white naturalists relegating those with darker skin to a lower status. It also draws out another issue: the lack of money in the mission. The strains that this put on John's research are a recurrent theme in his correspondence. On multiple occasions, he entreated his superiors to supply him with research funding.[131] One of the items on his list of desiderata was a colour-box. In a letter to Gottlieb Friedrich Stoppelberg (d. 1797), the man who oversaw the accounts of the *Franckesche Stiftungen*, John explained that he had thus far been unable to secure any good paints from Germany, and that he required an English paint cabinet costing around 30 Reichsthalers, which he thought both his pupils at the mission schools and his draftsmen could put to good use.[132]

Besides collecting and drawing, John was also dependent on the help of others for preserving his specimens. Where the previous chapters have already discussed the difficulties presented by the preservation of fishes, this was an even more trying undertaking the tropics. John describes how, while waiting for a ship to sail to Denmark to transport a freshly collected batch of fish in a cask, the rainy season had set in and the humid climate had destroyed his specimens, with worms appearing in them.[133] He then ordered fishermen to catch new specimens, let his European and Malabarian boys skin them, dried the skins in an oven and let them be coated by cajuput oil. Whenever possible he also sent an example of each species submerged in spirit so that Bloch could examine the internal anatomy of the fish. For this, glass jars needed to be filled with arak (a strong liquor made from juice of coconut trees, rice and sugar cane), which had to be refreshed multiple times. Needless to say, this process was both labour-intensive and expensive. John complained that he could scarcely get anythings that were required for preserving specimens such as appropriate knives and other such instruments necessary for stuffing animals, needles to pin up insects, glass jars and corks to close these with, cases to pack specimens in, as well as paper of adequate quality to preserve plants with. These materials were so scarce that he had to import from Halle at his

131 Liebau, *Cultural Encounters in India*, 264; in a response, Schultze explained that the natural historical works were too expensive to be sent, see: Schultze to John, 19 August 1790, AFSt/M 1 C 31b: 30.

132 John to Stoppelberg, 15 September 1791, AFSt/M 1 C 33a 87; to give some context, John's annual salary in the 1780s would have been 400 Reichsthalers; Liebau, *Cultural Encounters in India*, 193.

133 Christoph Samuel John, "Einige Nachrichten von der Küste Koromandel. Auszug eines Schreibens des Hrn Missionarius C.S. John an Hrn D. Bloch in Berlin," *Berlinische Monatschrift* 24 (1794): 357.

FIGURE 16 Specimen of *Salmo tumbil* from Bloch's collection. ZMB 32625, *Museum für Naturkunde*, Berlin

own expense. While John explained that he was eager to learn the *holländische Kunst* of stacking skins of fish amid wooden plates, he lacked the thin plates required for this method.[134] It is likely that he meant the ichthyarium method of Gronovius, which utilised such wooden plates.

At least eight of the fishes forwarded by John to Bloch are known to be still extant.[135] These include both wet and dry specimens and show a heterogeneity of preservation techniques. Fischer has shown, for example, how two examples of the same species of pufferfish might be preserved in radically different ways, each presenting an altogether different image of that species.[136] Earlier chapters have touched upon the considerable gap between a living animal and prepared specimen. One example of a dried specimen in his collection was a species of lizardfish, which Bloch named *Salmo tumbil* (Figure 16).[137] It is a stuffed exemplar, furnished with a layer of varnish, that displays an impressive row of teeth; its fins and their rays appear somewhat withered.

Another one of John's dried specimens is the carp that had been taken from the mission's pond (Figure 17). The Tamil called it the *Sölköndei* (the meaning of which is not explained in the description) and Bloch named it fringed carp

---

134 Ibid.

135 See: Paepke, *Bloch's Fish Collection in the Museum für Naturkunde*, 41–154.

136 Fischer, "The Afterlives of Fish Far from Home," 566.

137 Specimen of *Salmo tumbil* (currently accepted species name: *Saurida tumbil*) MfN, ZMB 32625. Described in *Allg. Nat. der Fische*, vol. 12, 112–113.

FIGURE 17 Specimen of *Cyprinus fimbriatus* from Bloch's collection. ZMB 8794, *Museum für Naturkunde*, Berlin

because of its fringed mouth.[138] In this particular instance, the carp's skin has been curved a little rather being pressed flat on the page, thus retaining some of the fish's original shape. While it may have been mounted onto a piece of wood made especially for this purpose to achieve this curve, no such cast is extant. A label, sadly illegible, has been pasted into the inner part of the skin. While this was a rather labour-intensive technique, it was also relatively cheap because less material was needed to make the specimen and it was easier to transport than glass. At the same time, the preservation method rendered the specimen somewhat fragile: fins might tear or break, which would hinder identification and classification. The work of cutting open fish to preserve their skin, turning their frail bodies into stable specimens came with challenges even under ideal conditions; doing so in the tropics required particular dedication and skill.

A cask of rotting fish, infested with worms. This is an image that undercuts Bruno Latour's notion of a 'centre of calculation', a centre where maps, specimens, diagrams etc. are accumulated and turned into universal knowledge so as to act at a distance.[139] Indeed, this idea has long since been complicated by historians, who have demonstrated that the control of those residing in

138 Specimen of *Cyprinus fimbriatus* (currently accepted species name: *Labeo fimbriatus*), MfN, ZMB 8794. Described in ibid., 50.

139 This concept is developed in Bruno Latour, *Science in Action: How to Follow Scientists and Engineers Through Society* (Cambridge Mass.: Harvard University, 1987), 215–257.

the centre of calculation was far from absolute.[140] Kapil Raj, for example, has stressed that new knowledge was created in contact zones where Europeans and South Asians interacted rather than in the more unidirectional movement Latour envisages.[141] In attempting to reconstruct the far-flung and multidirectional connections that made Bloch's collection possible, as this section has attempted to do, the limits of such a reconstruction have also become clear, in that local contributors to this collection remain only partly visible. We can also problematise the usefulness of the very notion of 'local', as it is often used by historians but seldom qualified. Who, precisely, do we see as local? Is it appropriate to speak of local in the context of the forced displacement of people? And is it imaginable that the missionaries, spending decades in their settlements, would at some point become local? Although there are perhaps no clear answers to such questions, it is worthwhile to consider them.

The (in)visibility of these collectors also has a visual expression. We find it within the vignette on the title page of the tenth part of Bloch's book series (Figure 18) set in a hilly landscape with a lake.[142] It is an allegorical, stylised depiction of the process of collecting. It portrays various putti engaged in the process of either packing or unpacking a wooden crate, the lid of which reads 'Tranquebar' – referring to the origin of many of the species described in the volume – and which contains fish submerged in glass jars. One of the putti examines a specimen while two others study a drawing of a fish. In the history of science, cupids have been seen as representations of invisible technicians, as Steven Shapin has argued for example for the seventeenth-century prints of the air pump of Otto von Guericke (1602–1686), where the instrument is operated by putti.[143] As Dominik Hünniger has argued, however, putti on frontispieces of natural historical publications convey both the manual and intellectual labour involved in the study of nature, as they are often depicted

140 Lissa Roberts, "Introduction: Centres and Cycles of Accumulation," in *Centres and Cycles of Accumulation in and Around the Netherlands During the Early Modern Period*, ed. Lissa Roberts (Berlin: LIT, 2011), 6.

141 See: Kapil Raj, *Relocating Modern Science: Circulation and the Construction of Scientific Knowledge in South Asia and Europe* (Basingstoke: Palgrave Macmillan, 2007).

142 Every volume of images carries its own vignette. Some of these are modelled after prints of Theodoor de Bry, published in his reprint of Thomas Hariot, *A Briefe and True Report of the New Found Land of Virginia* (Frankfurt: Johann Wechel, 1590). I thank Kim Sloan for pointing out this connection.

143 Shapin, "The Invisible Technician," 556; elsewhere, he and Schaffer have argued that the depiction of such cherubs is a standard convention in baroque illustrations, to imply that the depicted process of knowledge production was divine, see: Shapin and Schaffer, *Leviathan and the Air-Pump*, 334–335.

FIGURE 18 Vignette, Johann Carl Wilhelm Rosenberg (artist) and Daniel Berger (engraver). Marcus Élieser Bloch, *Allgemeine Naturgeschichte der Fische* vol. 10 (Berlin: J. Morino, 1793)
LEIDEN UNIVERSITY LIBRARY

observing and discerning species as well as gathering them.[144] The cupids on Bloch's vignette make visible the kinds of work entailed in the study of fish, stressing the collective nature of such an enterprise, while they render the workers themselves – John's largely unnamed collaborators and assistants – invisible. It is an example of a process of representing and effacing that we also see on other allegorical frontispecies.[145]

The fish collected on the Coromandel Coast included several species and even an entire genus that had not yet been described according to the Linnaean system. Bloch marked the latter feat by naming said genus *Johnfische*.[146] Moreover, John gained access to several communities of *Naturkenner*, both in Tranquebar as well as in Germany, through his natural historical efforts. By 1800, more than eight European learned societies had welcomed missionaries as

144 Dominik Hünniger, "Visible Labour? Productive Forces and Imaginaries of Participation in European Insect Studies, ca. 1680–1810," *Berichte zur Wissenschaftsgeschichte*, 44 (2021), 185.

145 Barnett, "Showing and Hiding," 248.

146 Or, in Latin, *Johnius*. *Allg. Nat. der Fische*, vol. 10, 132.

corresponding members.[147] John and Rottler, for example, were elected both to the Academy of Sciences Leopoldina, at that time based in Erlangen, and the *Gesellschaft Naturforschender Freunde* in Berlin. The names of missionary naturalists where thus codified in textual monuments such as scientific names or the membership lists of prestigious societies. The contributors to John's shipments of fish to Bloch, vital as they were to the natural histories that were being produced, would have no new species or genera named after them, however. And yet, the fish currently held in the *Museum für Naturkunde* and their paper counterparts in the books of Bloch are eloquent testimony to their work.

As the letter in the introduction to this chapter showed, the collectors of the fish, whom John referred to as "my black fishermen", might have seen the specimens that they had collected returned to them in illustrated form, swimming on the pages of Bloch's book. John had asked his superior to send him the newest volume of Bloch's book series because he only had an older, incomplete copy. As his requests to his superiors for materials could also fall on deaf ears, John was dependent on exchanges with European naturalists like Bloch and Chemnitz to procure such items as he needed for his studies. In a published letter, John thanks Bloch for sending him the latest volume.[148] The accumulation of a small library in Tranquebar was practical, because any questions that John received from his correspondents would often refer to published works such Bloch's.[149] Mostly, however, these books facilitated the process of collecting: natural historical books served as a list from which could be inferred those species still to be obtained, and those which were already identified.[150] The collection of fish in the colonial context of Coromandel was thus neither an isolated event nor a linear movement, but rather constituted a circular, iterative process to which John's many unnamed helpers proved vital.

4 To Capture Fishes on Paper

After the fringed carp that John's helpers had taken from the mission pond was converted into a specimen for Bloch's collection, it was described and depicted

147 Hanco Jürgens, "Van God's Akker tot Spiegel der Natuur: Veranderende Percepties van de Indiase Natuur in Berichten van Duitse Zendelingen," *De Achttiende Eeuw* 36, no. 2 (2004): 82.

148 John to Stoppelberg, 15 September 1791, AFSt/M 1 C 33a 87 and John, "Einige Nachrichten von der Küste Koromandel," 351; unfortunately, he does not mention which particular copy he has received.

149 Ruhland, *Pietistische Konkurrenz*, 337.

150 Anne Mariss, *A World of New Things: Praktiken der Naturgeschichte bei Johann Reinhold Forster* (Frankfurt am Main: Campus, 2015), 350.

in the last volume of the fish series.[151] Most of his specimens were subjected to a similar process. We will now take a closer look at the process of illustrating specimens and turn to the depiction strategies used by Bloch and his artists. Despite the centrality of illustrations to Bloch's project, these strategies have never been examined in detail before. The illustration process is richly documented compared to that other of many natural historical publications. This is, first of all, because the original drawings for a considerable proportion of the specimens in Bloch's collection are still extant, as are, of course the engravings subsequently published in the series – object, drawing and engraving can thus be compared. The second reason is that Bloch was quite explicit in formulating his ideas about what made a good illustration.

In this section, we analyse Bloch's illustration policies and discuss how he combined existing visual techniques with new ones to both present his fish in animated colours and to offer a three dimensional view of the specimen on paper. With the help of his artists, engravers, printers and colourists, he created 'epistemic images', serving not so much to only depict the object at hand but to replace it all together.[152] Bloch had those characteristics of the fish which he deemed essential recorded by his artists, the resulting illustrations becoming a substitute not of the preserved specimens in his collection, but the living, swimming species of fish these represented. They were nature revived. As we will see, Bloch and his various artists encountered both practical and epistemological challenges as they attempted to create impages that lived up to this intention.

Bloch outlined his illustration policy in the preface to the first volume of the book series. The three pages that he devotes to it show he had given a lot of thought to how one could produce the best visual representation of a species. A good drawing, in fact, began before lead or brush had touched the paper, by selecting the most suitable specimens. These were, according to Bloch, fresh as well as fully grown exemplars, because these best showed the species' distinguishing marks.[153] Bloch argued that such a meticulous approach to illustration was necessary because fishes tended to look alike, making it hard to distinguish one from the other.[154] He thus required his artists to be attentive to even the slightest deviations in a specimen. All details needed to be recorded on paper as they were relevant to get a proper understanding of the depicted species.

---

151 *Allg. Nat. der Fische*, vol. 12, 50.

152 Daston, "Epistemic Images," 17.

153 *Allg. Nat. der Fische*, vol. 1, sig. *3r.

154 Ibid., sig. *3v.

Over the more than a decade that it took to publish his book series, Bloch employed no fewer than nineteen draughtsmen and engravers.[155] The signature of the painter Johann Friedrich August Krüger (b.1754) appears on many on the plates; he was also commissioned by Martini for a series on shells.[156] The drawings made by draughtsmen were turned into engravings by, among others, Johann Friedrich Hennig (b.1778), Johann Godlieb Schmidt (1750–1822), and Georg Bodenehr (dates unknown). It remains unclear who subsequently coloured these engravings. At the time, it was common to hire women or children for this part of the process, because they were paid less even if the work they performed required considerable skill.[157]

Bloch highlighted six further areas to which contributing artists were to pay particular attention. First of all, he stated, they needed to convey the proportions of the specimen properly. Secondly, they had to represent the position and the shape of the fins correctly, particularly that of the caudal fin; these were, after all, important marks for classification. Thirdly, the precise number of bones in the gill flap as well as the number of rays in the fin were to be clearly represented. Bloch explained the reason that both were to be delineated in the same terms as Artedi, noting that the former was necessary for deciding genera, and the latter for deciding species.[158] The fourth rule was to give a truthful representation direction of the lateral line (the thin line on the side of the fish stretching from its head to its tail).[159] The fifth rule was that the artist had to consider several different elements to ensure they showed the fish's scales accurately: their size, placement, and shape as well as any pattern of stripes or dots they might display.[160] The sixth and last item on the list of instructions was that artists should always include the 'natural colour' of the fish.[161]

These were the pictorial ideals that needed to be put into practice. The process began with the creation of a drawing of a specimen in preparation for the engraving. Fortuitously, some two hundred of these of *Originalzeichnungen* still exist in the *Historische Arbeitstelle* of the *Museum für Naturkunde*.[162] These

155 Few of the names of these artists are known; Wells, "M.E. Bloch's Allgemeine Naturgeschichte der Fische," 9.

156 Claus Nissen, *Die Zoologische Buch-Illustration*, vol. 2 (Stuttgart: Anton Hiersemann, 1978), 153 and Wells, "M.E. Bloch's Allgemeine Naturgeschichte der Fische," 9–10.

157 Nickelsen, *Draughtsmen, Botanists and Nature*, 62.

158 Ibid.

159 Ibid., sig. *4r.

160 Ibid.

161 Ibid.

162 Bound volumes with drawings of fish for plates CCI to CCCCDDDI (with some gaps), ZMB, VIII/423 and VIII/424. It seems that there were originally two more of these volumes, according to Karrer, "Marcus Elieser Bloch," 145.

drawings are assembled together in two volumes, with many of them being carefully pasted onto the bound pages.[163] Each drawing has been executed in colour – sometimes in watercolour, other times in what appears to be a thicker, gouache-like paint. Let us take a look at one particular drawing: the one that Johan Friedrich Hennig made for the lizardfish that John had sent to Bloch (Figure 19). The drawn fish corresponds to the preserved specimen in its shape (although the depicted specimen is somewhat more plump) and its open mouth displaying rows of teeth. While the position of the fins in the drawing correspond to their placement on the specimen, their aspect does not: while the fins on the specimen have dried out, the drawings show them fanned out, as they would have seemed when the fish had still been under water. The most striking difference between specimen and drawing, however, is to be found in the palette of colours used by the colourist. As no traces of colour remained in the specimen, and John's rendition of its colours seemed similarly bereft, Hennig was put in a delicate position when deciding the appropriate colours. He opted for a brownish, grey colour with maroon overtones. As we will discuss shortly, the engraving based on this drawing would have a different colouration (Figure 20).

The drawings were subject to revision. Given Bloch's very clear opinions on the subject, it seems unlikely that a drawing would be sent to the engraver without his stamp of approval. It was common for naturalists to closely supervise their draughtsmen, making sure that they drew the relevant, correct features.[164] A symbiotic collaboration between artist and naturalist with regard to drawing would result in, as Lorraine Daston and Peter Galison have put it, 'four-eyed sight'; the head of the naturalist fusing with the hand of the artist.[165] That there must have been at least some discussion between Bloch and his artists is evidenced by the drawings themselves. A few of them contain corrections in graphite or sometimes ink.[166] In some instances, Bloch or one of his draughtsmen might alter a fish's shape. In one image, a graphical edit was suggested, indicating that the fish's eye might be placed more accurately.[167] Clear evidence of the assertion of Bloch's third rule, which focused on the correct representation of the number of rays in the fins, can be found in the annotations included in certain drawings besides each of the fins of the fish that indicate how many rays it has, and how many of these rays are spikey

---

163 A practice also used in the drawings of *aquatilia* in the Gessner-Platter albums mentioned in Chapter 1 and 2.

164 See, for example: Fransen, "Antoni van Leeuwenhoek, His Images and Draughtsmen," 493.

165 Lorraine Daston and Peter Galison, *Objectivity* (New York: Zone Books, 2007), 88.

166 Karrer, "Marcus Elieser Bloch," 146.

167 Drawing of *Chaetodon nigricans* (currently accepted species name: *Acanthurus nigricans*), ZMB, VIII / 423, 3.

FIGURE 19 Original drawing of *Salmo tumbil*, Johann Friedrich Hennig. ZM B VIII / 423, Bl. 430, *Archiv Museum für Naturkunde*, Berlin

(Figure 21).[168] Bloch wanted his illustrations to be precise enough for naturalists to be able to identify and classify any fish based on its image alone where possible. His intention was thus very much the opposite of that Artedi, whose descriptions rendered drawings superfluous. Bloch's wish to cater to a broader audience than a literate or Latinate one may have been one of the reasons why he considered images so important.

After drawings had been approved, the next step was to engrave them. Although the engravings do, as one can reasonably expect, differ somewhat in colouration between physical copies of Bloch's works, overall they are chiefly consistent especially with regard to the considerable intensity of the paint.[169] Overall, the engravings follow the model drawings closely, in everything except colour.[170] The colours of the original drawings seem more subdued that those of their engraved counterparts. This might be a result of the specific pigments or paints used, but may also have to do with whether or not a layer varnish was applied, as well as the mode of storage used; lengthy or recurrent exposition to sunlight would cause the colours to fade. Although it is of course impossible to ascertain the extent to which the colours of these engravings and drawings

168 Drawing of *Bodianus maculatus* (currently accepted species name: *Plectropomus maculatus*), ZMB B VIII / 424, 21.

169 I base this statement on having perused physical copies in Leiden University Library, Artis Library and Teylers Museum alongside digitised copies at Biodiversity Heritage Library (https://www.biodiversitylibrary.org, last accessed 15 February 2025).

170 Karrer, "Marcus Elieser Bloch," 146.

FIGURE 20 Engraving of *Salmo tumbil*, Johann Friedrich Hennig. Marcus Élieser Bloch, *Allgemeine Naturgeschichte der Fische*, vol. 12 (Berlin: J. Morino, 1795), plate CCCCXXX
LEIDEN UNIVERSITY LIBRARY

FIGURE 21 Original drawing of *Bodianus maculatus*, Johann Friedrich Hennig. ZM B VIII / 423, *Archiv Museum für Naturkunde*, Berlin

have altered over the past centuries, however subtly, these differences between drawings and engravings seem significant considering the very specific demands placed by Bloch on replicating the natural colour of the specimen.

Naturalists and the artists whom they employed were well aware that time was of the essence in illustrating fish. They started drawing the specimen as soon as possible after it was caught, and developed several tricks to aid naturalistic rendition. In the preface to his natural history of Carolina, Catesby (whose pictures, we read in the previous chapter, Gronovius deemed to be painted very well) explained that: "fish, which do not retain their colours when out of their element, I painted at different times, having a succession of them procur'd while the former lost their colours."[171] Ferdinand Lukas Bauer (*c.*1760–1826), one of the artists accompanying the naturalist John Sibthorp (1758–1796) on his expedition to the Mediterreanan, deployed a sort of painting-by-numbers technique which allowed him to indicate the shades of colour of specimens without having to actually paint them on the spot.[172] Although Artedi and

171 Mark Catesby, *Natural History of Carolina, Florida and the Bahama Islands* (London: W. Innys and R. Manby, 1729–1747), xi.

172 Rees, *Die Verzeichnete Fremde*, 199.

Linnaeus considered colour to be of secondary importance, other naturalists and their artists continued to put considerable time and effort into capturing it accurately.

Although the sixth rule, which specified that artists should portray the natural colour of the fish was quite straightforward in principle, obeying it in practice was not quite so simple. First of all, it could be a tricky exercise to exactly replicate the natural colour of a fish. In his description of the golden tench, a gift from the palace pond of Elisabeth Christine, Queen of Prussia (1715–1797), Bloch lamented the fact that the artists, despite their substantial skill and best efforts, had not managed to truly capture the beauty of the specimens' natural colour.[173] Of course, such a response would have been entirely appropriate considering the effectively royal status of the fish as gift, but it also exposes a certain tension underlying the work of colouring fishes. Bloch could of course only have assessed whether colours had been mixed correctly when he had a living specimen at hand for comparison. That was not the case for any fish that was sent to him by a correspondent.

In the case of the aforementioned fringed carp, its species description reveals that the specimen had been accompanied by a drawing from the life on which Bloch's draughtsmen could base their engraving of the fish; as mentioned, however, not one of the illustrations sent over by John has survived.[174] Even in those cases where John's drawings are mentioned in a species description, no indication is made of whether or not they included colour; recall John's difficulties in securing proper paint. In the cases such as the lizardfish, where no accompanying drawing is mentioned, Bloch's draughtsmen must have had to make do with the preserved objects themselves and the verbal descriptions that possibly accompanied them. Exactly how draughtsmen and colourists went about the business of replicating the exact colourings of any individual fish when all they had to go one was an essentially monochrome preserved specimen (and perhaps some verbal description) remains somewhat of a mystery.

Colour was applied both during the printing process and after. As discussed in Chapter 1, early modern printed works occasionally included colour: this could be due to the use of coloured printing ink, but in most cases, colour was

173 "Ich muss bekennen, dass, ohnerachtet die Künstler bei dem Ausmalen desselben allen Fleiss angewendet, sie doch noch weit zurück geblieben sind, die Schönheit der natürlichen Farben zu erreichen." *Allg. Nat. der Fische*, vol. 1, 90. Currently accepted species name: *Tinca tinca*.

174 Karrer, "Marcus Elieser Bloch," 146.

FIGURE 22 Engraving of *Scomber ruber*, Johann Friedrich Hennig. Marcus Élieser Bloch, *Allgemeine Naturgeschichte der Fische*, vol. 10 (Berlin: J. Morino, 1793), plate CCCXLII
LEIDEN UNIVERSITY LIBRARY

FIGURE 23 Original drawing of *Scomber ruber*, Johann Friedrich Hennig. ZM B VIII / 424, Bl. 342. *Archiv Museum für Naturkunde*, Berlin

applied by hand.[175] In Bloch's series of fish books, however, we find a unique combination of both types of colour administration. Mechanical colour printing had by this time become a more established technique.[176] Although most of the plates are printed in black or grey ink, some 10% of the plates were printed in brown, orange, red, and green.[177] Colour printing was still relatively crude, however, and not suitable for displaying sophisticated gradations and variegations. This is why all Bloch's prints were also coloured by hand. The function and effect of colour printing can be seen on the engraving of the red mackerel (Figure 22). For this fish, a striking red has been used – as discussed earlier, Bloch had stipulated that its red colour set this species apart from other mackerel.[178] The ink used for the printing of the plate is red.[179] This gave the engraved illustration the correct undertone, on which the colourists could layer their own colouration.[180] The effect is an intense red colour, which

175 See also: Kusukawa, *Picturing the Book of Nature*, 69–81.

176 For the historical development of colour printing, see Elizabeth Savage and Ad Stijnman, "'Material Colour': The Heritage of Colour Knowledge in Seventeenth- and Eighteenth-Century Printshops," in *Colour Histories: Science, Art, and Technology in the 17th and 18th Centuries*, eds. Magdalena Bushart and Friedrich Steinle (Berlin: De Gruyter, 2015), 95–113 and Margócsy, *Commercial Visions*, chapter 4 "Knowledge as Commodity: The Invention of Color Printing," 167–199.

177 43 of the total of 432 plates use coloured ink rather than black. I have counted the illustrations of the copy at UBL, 137 A 1/6, but have not been able to verify these exact numbers for other copies.

178 *Allg. Nat. der Fische*, vol. 10, 75.

179 Engraving of *Scomber ruber*, *Allg. Nat. der Fische*, plate CCCXLII.

180 I am indebted to Sabine Hackethal for this insight.

is especially striking when compared to the subtle tints of the original drawing made for this species (Figure 23).[181] A similar contrast can be seen in the case of the lizardfish, where the original drawing (Figure 19) offers distinctly subdued colouration when compared with the published engraving (Figure 20).

At the same time, one has to be careful in assuming what was and what was not known about the actual colourings of specimens. Although Claudia Kreklau has argued, for example, that the dried exemplars of foreign fishes, with their brown hues, "fit comfortably in the naturalists' worldview of a dark and dreary underwater world",[182] it appears that Bloch, at least, was not prejudiced by such a dismal outlook. Bloch repeatedly marvelled at the splendid colours displayed by the fishes of faraway regions. In the preface to the ninth volume, for example, he exclaimed that its plates dedicated to foreign fishes distinguished themselves from those made after German fishes by their beautiful colours.[183] He confidently stated this without ever having seen a living fish from, say, the East Indies in the flesh. Bloch's statement fits into a broader exoticist discourse, which lauded the spectacular colours of plants and animals in the warmer climes.[184] As Kreklau seems to overlook, while stay-at-home naturalists like Bloch could not glimpse the original, dazzling colours from preserved specimens, they may well have had access to coloured drawings of either these or similar species, as well as colourful descriptions of them in letters or other textual works. These second or even third-hand colourisations of foreign fishes would become part of the visual repertoire of both naturalists and their artists, and would literally colour any subsequent encounters with specimens from a similar 'exotic' origin.

Bloch purposefully sought out other illustrations of fishes. He consulted, for example, a copy of the vividly, if somewhat fantastically, coloured *Poissons, écrevisses et crabes de diverses couleurs et figures extraordinaires* (1719) by Louis Renard.[185] At a public auction in Berlin, furthermore, he obtained a collection of drawings of Caribbean fishes by the French missionary Charles Plumier (1646–1704).[186] He also visited the city's Royal Library to peruse manuscript volumes containing drawings of fish from the West Indies, made under the auspices of Johan Maurits van Nassau-Siegen – an expedition discussed in

181 Drawing of *Scomber ruber*, ZMB B VIII / 424, 342.
182 Kreklau, "Travel, Technology, and Theory," 596.
183 *Allg. Nat. der Fische*, vol. 9 (Berlin: J. Morino, 1792), sig *2r.
184 Schmidt, *Inventing Exoticism*, 11–12.
185 Theodore W. Pietsch and Justin R. Hanisch, "Louis Renard (1678/1679–1746) and His *Poissons, ecrevisses et crabes* (1719): 300 Years of One of Natural History's Most Curious Colour Plate Books," in Smith and Egmond, *Ichthyology in Context*, 584.
186 Theodore W. Pietsch, *Charles Plumier (1646–1704) and His Drawings of French and Caribbean Fishes* (Paris: Publications Scientifiques du Muséum, 2017), 83.

Chapter 1.[187] While Bloch prided himself on offering original illustrations in his works, created under his direction in his own studio after fresh or preserved samples of fish, he did have his artists copy drawings from the abovementioned manuscripts. As he explained in the preface of one of his volumes, "both [authors] had faithfully depicted the fish on the spot and painted them in lively colours."[188] That the drawings had been made *ad vivum* made them reliable enough to merit inclusion in Bloch's own work. Chapter 1 has shown that it was, in fact, quite common to use one illustration as a reference drawing for the other. Nickelsen has argued for such 'copying links' that they are not merely derivative drawings, but the outcome of a conscious process of including and excluding certain elements of the initial illustrations.[189]

One of the most striking aspects of Bloch's illustrations is that a considerable proportion of them have been heightened in silver and gold. This means that silver and gold paint is applied to certain parts of the fish, for example its scales and certain parts of its head; just as Baldner's artists had done with white in Chapter 2. Substituting white with gold and silver was not a completely new practice, but was still altogether rare.[190] The paints were, in all likelihood, made by mixing powdered silver or gold leaf with a binding medium such as gum arabic.[191] This was an expensive procedure, and one that an eighteenth-century painting manual therefore advised should only be used for special occasions.[192] Bloch seemed to believe that these costs were warranted. One reason for this might have been the sense of luxury that it added to his series of fish books. Another reason may have been the particular effect that it produced. This has to do with the luminescent qualities of silver and gold: the metals in the paint capture and reflect light. In applying the paint judiciously to the parts of a fish that would naturally catch the light, such as the edges of

187 This material is discussed in more detail in Whitehead and Boeseman, *A Portrait of Dutch 17th Century Brazil*, 40–43.

188 "[…] beide haben die Fische an Ort und Stelle getreu abgebildet, und nach lebendigen Farben ausgemalt." *Allg. Nat. der Fische*, vol. 6, sig. a2r.

189 Nickelsen, *Draughtsmen, Botanists and Nature*, 203–214.

190 An early example of applying gold to natural historical drawings of fish is John White (1539–1593). See: Kim Sloan, *A New World: England's First View of America* (Chapel Hill: University of North Carolina Press, 2007).

191 This technique was also known as 'shell gold', because the ingredients were often mixed in shells. See also: Michèle Seehafer, "Shimmering Virtue: Joris Hoefnagel and the Uses of Shell Gold in the Early Modern Period," in *Materialized Identities: Objects, Affects, and Effects in Early Modern Culture, 1450–1750*, eds. Susanna Burghartz, Lucas Burkart, Christine Göttler and Ulinka Rublack (Amsterdam: Amsterdam University Press, 2021), 281–321.

192 Willem Goeree, *Verligterie-kunde, of regt gebruik der water-verwen* (Amsterdam: Andries van Damme, 1705), 22.

its scales and its gill cover, the artist endowed the illustration with some of the vivacity, as the silver or gold pigments imitated the glistening sunlight on the wet scales of the fish as it is plucked from the water. Others sought to replicate this technique. When requesting the purchase of a colour-box for his draughtsmen, John added that he also lacked sufficient silver and gold colour for them to draw the fishes.[193]

Another remarkable aspect of Bloch's fish illustrations is the degree to which they consider the fact that the specimens which they reproduce exist in three dimensions. The majority of plates contain a representation of the fish in cross section. This visual technique was used by William Hooke in his seventeenth century depictions of ammonites,[194] while the aforementioned Plumier offered cross-sectional views with his drawings of Caribbean fish, of which Bloch owned a manuscript. Bloch was the first to consistently include such sections for fish in a printed work.[195] An example can be seen on the engraving of the red mackerel (Figure 21). It is a schematic depiction of what you would see if you were to cut the fish in half at its thickest point. In his illustration policy, Bloch explains, in one sentence, the purpose of these contours: to offer an idea of whether a species was thick or thin.[196] That cannot have been the sole intent, however. For even though Bloch does not mention it, besides outlining the circumference of the fish's body, the sections also indicate the shape and location of its spine and abdominal cavity.[197] The dissection of a fish, which as we saw in Chapter 1 could be a rather complex process, revealed those parts not usually visible. The technique of the cross section offered a neat, abstracted representation, and was more widely used to portray the properties of both human and animal tissue.[198] To create such a cross section for fish required the

193 "An einen guten Quantität Silberfarbe und etwas Goldfarben fische zu zeichnen fehlt es mir auch sehr eigentlich an feine Pinseln." John to Stoppelberg, 15 September 1791, AFSt/M 1 C 33a : 87.

194 Sachiko Kusukawa, "Drawings of Fossils by Robert Hooke and Richard Waller," *Notes and Records of the Royal Society of London* 67, no. 2 (2013): 124.

195 These cross sections also appear in later works of the 1780s, for example Pierre Joseph Bonnaterre's *Tableau encyclopédique et méthodique des trois règnes de la nature: Ichthyologie* (Paris: Mme Veuve Agasse, 1788). Bonnaterre also prepared the fish illustrations in Diderot's *Encyclopédie* depicting similar sections.

196 "Damit man aber auch wissen möge, ob der Fisch dick oder dünn ist; so habe ich einen Umriss van stärksten Theile deselben beygefügt." *Allg. Nat. der Fische*, vol. 1, sig. *4r.

197 The cross section also offered an indication on both the muscle mass and the amount of meat of the fish – possibly useful knowledge for consumption. I thank Martien van Oijen for this insight.

198 For examples, see: Domenico Bertolini Meli, *Visualizing Disease: The Art and History of Pathological Illustrations* (Chicago: Chicago University Press, 2017), 79 and Matthias Bruhn, "Beyond the Icons of Knowledge: Artistic Styles and the Art History of Scientific Imagery,"

cutting of at least one specimen in half – the specimen would then, of course, no longer be intact. This might explain why almost all of the plates of German fish, but only around half of the foreign ones, of which Bloch could not easily attain a wide sample of each species, include this cross section. It also required a complete specimen: making such a cross section was not possible from dried exemplars, from which all flesh and intestines had already been removed.

Besides showing the thickness of species on the engraved plates, Bloch also indicated their magnitude. Many of the plates indicate whether the engraved image reflected the species' true size [*natürliche Grösse*], or whether it offered a reduced [*verkleinerte*] view.[199] This attention to the actual shape and size of a specimen was of course far from new in natural historical depictions. The representation of the 'true size' of observed entities had been a pressing problem especially for early microscopists of the seventeenth century, who often included illustrations of the naked eye view alongside the magnified drawing one to indicate the scale.[200] Bloch's specimens were not microscopic entities, however, and his size indications are less exact in nature: his aim was only to indicate that a specimen was either larger or smaller than the illustration made from it. Taken together, the decisions to include cross sections of and size indications for the specimen helped the onlooker to envision the general size and shape of the actual, three-dimensional object on which the engraving was based from its two-dimensional representation on the page.

In sum, the coloured engravings were the result of carefully measured decisions on part of Bloch and his artists. The artists set to work capturing the properties of the specimens on paper in detail – filling in the gaps where necessary. As we saw with the lizardfish from Tranquebar, a specimen and a drawing of an individual species might differ markedly, especially with regard to shape and colour, while the metamorphosis from drawing to engraving was most likely to affect a fish's colouration. Certain features could be altered in this three-stage act of preservation, while others might be lost irrevocably. Even though some naturalists considered illustrations to be suitable substitutes for objects, they were mediated by necessity. It is not now possible to recover the full extent of this mediation: historians cannot hope to determine just how much the physical state of these materials has changed over the past two centuries, to what degree the specimens, drawings and engravings look the same as they did in the eighteenth century.

---

in *The Technical Image: A History of Styles in Scientific Imagery*, eds. Horst Bredekamp, Vera Dünkel and Birgit Schneider (Chicago: Chicago University Press, 2015), 41.

199 *Allg. Nat. der Fische*, vol. 1, sig. *4r.

200 Fransen, "Antoni van Leeuwenhoek, His Images and Draughtsmen," 506–509.

Bloch was aware that even the most truthful (*allergetreuesten*) illustrations could not always express all those marks that were of pertinence for classification, as for some species of fish these were located inside the body.[201] While the illustrations thus might help in classification, that was not their only function. To understand why Bloch considered the time and costs incurred in these efforts worthwhile, the notion of a 'paper museum' can be useful. Historians have used this term to describe various paper formalisations of collections from the seventeenth up to the nineteenth century.[202] Debora Meijers has succinctly defined it as "a group of drawings whose coherency stems from a deliberate effort of conservation."[203] Bloch's series of fish books not only preserved the fishes in his collection in their splendid, living state (albeit in mediated form), it also preserved the collection as a whole. Just like specimens themselves, after all, natural historical collections conceived as a whole were also in danger of disintegration. The fate of Seba's collection is a case in point. While his first collection had been purchased by Peter the Great, the apothecary's death led to his second collection being auctioned off piecemeal, its specimens dispersed into the collections of a wide variety of naturalists.[204] Many other collections suffered a similar (or worse) fate. The quarto volumes bearing Seba's name, however, with its coloured illustrations of the objects accompanied by brief descriptions, outlived his physical collection and, in a very real sense, preserved it much as he had preserved its individual specimens in the first place. Creating a visual record of the objects in one's collection brought particular advantages. A paper museum that displayed the plants in a botanical garden, for example, as Meijers explains, "showed them in their unchanging beauty, alive and intact, and blooming all at the same time."[205]

---

201 *Allg. Nat. der Fische*, vol. 10, sig. 2r.

202 Francis Haskell and Henrietta McBurney, "The Paper Museum of Cassiano dal Pozzo," *Visual Resources* 14, no. 1 (1998): 1–17; Debora J. Meijers, "The Paper Museum as a Genre: The Corpus of Drawings in St Petersburg within a European Perspective," in *The Paper Museum of the Academy of Sciences in St Petersburg c. 1725–1760*, eds. Renée Kistemaker, Natalya Kopaneva, Debora J. Meijers and Georgy Vilinbakhof (Amsterdam: Royal Netherlands Academy of Arts and Sciences, 2005), 19–54; Martin Rudwick, "Georges Cuvier's Paper Museum of Fossil Bones," *Archives of Natural History* 27, no. 1 (2010): 51–68.

203 She distinguishes between "real" paper museums, which are representations of objects existing in an actual collection, and "wish-list" paper museums in which drawings act as substitutes for objects not in the possession of the collector. Meijers, "The Paper Museum as a Genre," 25.

204 The fate of the objects of Seba's second collection has been traced in Boeseman, "The Vicissitudes and Dispersal of Albertus Seba's Zoological Specimens," *Zoologische Mededelingen* 44, no. 13 (1970): 177–210.

205 Meijers, "The Paper Museum as a Genre," 33.

As the discussion of the difficulties of making drawings from specimens has highlighted, it is important not to assume the (perceived) vicarious quality of drawings. While Martin Rudwick has contended that drawings of fossils could act as a stand-in for actual specimens, he attaches the caveat that "an effective proxy experience was necessarily mediated by the social and artistic conventions that underlay any pictorial representations in a given historical and cultural context."[206] This process of mediation, as has been discussed earlier in this chapter, obscured some parts of the depicted object, highlighted others, and might even add something new (whether wittingly or unwittingly). It is not difficult to imagine that the creation of a paper museum held an appeal to Bloch, as it both preserved the collection itself and made its general circulation possible. The fish specimens on the shelves in Berlin were thus only one manifestation of his collection – the book series was another.

As has been stressed above, Bloch's project was iterative in nature. As soon as a volume had found its way into bookshops and from there into the libraries of barons, countesses, physicians, merchants and naturalists, it invited its readers to send ever more specimens Bloch's way. As we saw, at least some of the volumes travelled to John in Malabar. In the *Auftausch* between Bloch and John of books and specimens, the series' illustrations played an essential role. While their size and cost made taking Bloch's volumes into the field somewhat impractical, John appears to have used the illustrations to communicate with those of his collectors who could not read the descriptions. As he explained, the coloured engravings allowed for "every dumb fisherman and journeyman" to collect the desired specimens.[207]

Together with his artists, Bloch developed an innovative pictorial format which helped bring the fish alive on the pages of his books. With his carefully developed and executed format for depicting fish, Bloch seems to have set a new standard. We see that, in the nineteenth century, images remained an essential component in natural historical publications of fish, such as the *Histoire naturelle des poissons* [Natural history of fishes] by Georges Cuvier and Achille Valenciennes and the *Atlas ichthyologique des Indes Orientales Nééerlandaises* [Ichthyological atlas of the Dutch East Indies] of Pieter Bleeker.[208] As is the

206 Martin Rudwick, *Bursting the Limits of Time: The Reconstruction of Geohistory in the Age of Revolution* (Chicago: Chicago University Press, 2005), 76.

207 "Hat man Werke mit gemahlten Zeichnungen, so kann jede dumme Fischer und Tagelöhner samlen halfen wie ich bey Chemnitz Conchylien Werk und Blochs Naturgeschichte der Fische ofte erfahre." John to Schulze, 20 January 1789, AFSt/M 1 C 30c : 24.

208 Georges Cuvier and Achille Valenciennes, *Histoire naturelle des poissons*, 22 vols. (Paris & Strasbourg: chez F.G. Levrault, 1828–1849); Pieter Bleeker, *Atlas ichthyologique des Indes Orientales Nééerlandaises, publié sous les auspices du Gouvernement Colonial Nééerlandais*, 9 vols. (Amsterdam: Frédéric Muller, 1862–1878).

case in Bloch's series of books, these works feature fish that are depicted on full folio plates, carefully coloured, often containing a cross-section of the species or an important detail of its anatomy highlighted. Just as in previous centuries, however, the nineteenth century saw the precise use and function of images in the natural history of fish continue to be hotly debated, not least the perennial question of how to preserve their colour on paper.[209]

The coloured plates fulfilled different functions. For Bloch, the book series may have served as a paper museum to his collection, the illustrated fish far surpassing the beauty of that of the specimens stowed away on the shelves of his cabinet. It is conceivable that a part of his audience was primarily attracted to the luxuriously executed plates. For naturalists, the carefully executed illustrations were a useful work of reference because of the care that had been taken to depict those characteristics that were salient for classification, for example correctly portraying the number of rays in the fins. As we saw, John used the illustrations as a way to communicate to his collectors which species still needed to be gathered, and so the plates had mnemonic value. The high selling price of these volumes, however, meant that Bloch's target audience of fishermen and farmers were probably unable to actually afford the volumes. Considering the purposes these coloured engravings actually served, as simultaneously beautiful and functional images, allows for a better appreciation of the ways in which these illustrations acted even beyond Bloch's own intentions for them.

5 Conclusion

In the eighteenth-century German states, the study of nature was a popular pastime. It attracted a varied group of people, including merchants, physicians, professors, apothecaries, book printers, agriculturists, among others. In the cameralist fashion of the time, investigations into living nature tended to focus on the ways in which the extraction of minerals, the cultivation of plants and the breeding of animals might advance the general health and wealth of the (envisioned) nation. Such pursuits were often presented as a patriotic act, though this was, of course, dependent on their not being followed for commercial gain. Fish also had their part to play in this shared project to chart nature along increasingly nationalistic lines, even though in following their own migration routes they slipped effortlessly beneath the borders that people had so fervently constructed. Bloch initially presented his work as an oeconomic natural history of fishes of the German states. By providing carefully coloured

209 See: Robbert Striekwold, "Images, Specimens, and Species: Hermann Schlegel on the Various Ways of Depicting a Fish," in Smith and Egmond, *Ichthyology in Context*, 701–726.

illustrations, he hoped that anyone might easily recognise the fishes that he described in his books, and in doing so might learn to see and use these fish as natural resources. Overall, Bloch's series was celebrated as an accomplishment for and of Germany, the characterisations of his work by contemporaries often striking a tone of national pride.

As this chapter has suggested, by portraying fish Bloch also presented himself. His collection and series of fish books were closely connected to his reputation as a *Naturkenner* in Berlin, the German states and in the wider enlightened circles of Europe. One of Bloch's eulogists called his fish oeuvre "the crowning glory that placed him among the ranks of our illustrious naturalists."[210]

With his book series, Bloch added no fewer than 250 new species to the European natural historical canon. As a careful compiler with an impressive library, he corrected existing descriptions of fishes where necessary, comparing and contrasting different accounts and by drawing on the specimens in his collections. What distinguished this book series apart from earlier encyclopaedic books of fish, of course, were the more than 400 sumptuously executed plates.

While it is unlikely that Bloch made any financial profit from his series – as can be inferred from his periodic references to the costs and expenses he incurred – he did have something to gain. As we saw, as a Jew in Berlin he was subject to several forms of exclusion. Firmly embedded in a group of Jewish enlightened thinkers, he also sought to establish connections with other *Naturkenner* in Berlin and beyond. In this, the assembly of a cabinet of *naturalia* was paramount. Through expanding his collection and publishing his series of fish books, he created a vast network. This network can be partially reconstructed from the names mentioned in the books' species descriptions, its subscription lists and sponsored engraved plates, and his correspondence. Joseph II and other monarchs expressed their interest in and support of the project, as did counts, bankers, book sellers, professors, missionaries, surgeons, and apothecaries, as well as a fishing guild. Up until the moment of the publication of the very last volume of his series, Bloch's books and collection reinforced one another – the publication of this series of fish books was a dynamic rather than a static, linear process. In this chain of collectors spanning different continents, Bloch was positioned at the end.

In considering natural historical works published by naturalists from countries without a continuous presence outside of Europe, one can easily miss the ways in which these naturalists did, in fact, benefit from colonial

210 "Mais son immortel ouvrage sur l'ichthyologie acheva de mettre le comble à sa gloire, en le plaçant au rang de nos illustres naturalistes." Antoine-Jean Coquebert de Montbret, "Éloge de Monsieur Bloch," *Rapport général des travaux de la Société philomatique de Paris* 4 (1800): 145.

infrastructures. Bloch was only able to produce his series of fish books because of the many shipping routes that had been created by the expansion of trade and religion in the preceding centuries. The Danish-Halle mission was of particular importance to Bloch's project, as it formed an important link between Germany and India. From his post in South India, Christoph Samuel John sought to align himself with Bloch's project, providing Bloch with no fewer than 50 fishes as he did so. By inserting himself into Bloch's network in this manner, the Pietist missionary also furthered his own identity as a naturalist participating in learned circles in Europe. Just as Bloch relied on men such as John, so the missionary, in his turn, depended on his own network of contributors, which included his European and Tamil pupils as well as fishermen and cooks. These individuals played a central role in every stage, from collecting to preserving, from describing to illustrating, and finally in packing everything for transport to Berlin. These contributors, however, are only partly visible. The language used by historical actors such as Bloch and John to describe these contributors – language often unwittingly echoed by contemporary historians when they use terms such as 'local' without qualification – has tended to obscure their identities.

While Bloch continued the classificatory approach that Artedi and Linnaeus had promulgated in their works, he conceived of his own work as an improvement of theirs. With its combination of elaborate descriptions and precise illustrations, Bloch's work simplified the often thorny and contentious business of identifying a species. The illustrations themselves were based on the preserved specimens of fish in his own collections – specimens which had already been selectively edited. Drawing on established techniques and introducing new ones, Bloch succeeded in developing a format for his 'epistemic images' that were to act as enhanced substitutes for his specimens. He introduced, for example, the use of cross-sectional views of specimens in print, experimented with innovative techniques such as the mechanical printing of colour, and heightened some of his illustrations with paint made from silver and gold. These decisions served to present the fish as three-dimensional, living beings that were revived on the pages of Bloch's works. His work set a new standard for the depiction of fish that would be emulated in later ichthyological publications.

As we saw, Bloch's distance from living, foreign fish that he and his artists only saw as preserved specimens created epistemological tensions, especially given his own extensive rules for how a fish should be represented on paper. Where the actual process of converting these specimens onto paper remains somewhat of a mystery, we can consider the effect of these engravings. Nevertheless, the pens and brushes of the draughtsmen and colourists rendered these diverse fish, caught and collected by many individuals from various regions, in a uniform style. Any epistemological uncertainties caused by the interpretation of the colours and other characteristics from dried specimens rather than living, breathing fish were not expressed. There was, also, power in repetition. Under the hands of the artists, the engraved plates were presented as a continuous set, painted in the animated colours of life. As Bloch could claim first-hand knowledge of many of the German fishes, he could, by association, allege a similar experience for the foreign fish. Bloch became a recognised authority on all the world's known fish without having to travel far from his Berlin home, let alone leaving Europe.

By the time he passed away in 1799, Bloch had become a celebrated figure in a country that nevertheless refused to grant him full legal equality. Friedländer was involved in handling his legacy, which entailed finding a fitting home for the collection.[211] Along with a dozen or so members of the *Gesellschaft Naturforschender Freunde*, he petitioned King Friedrich Wilhelm III (1770–1840) to purchase Bloch's cabinet.[212] Friedrich Wilhelm was in the process of turning his *Kunstkammer*, which comprised coins, antiquities, ethnographic objects and naturalia, into Berlin's first public museum.[213] In their letter to the king's representatives, the *Gesellschaft* members argued that Bloch's cabinet, and specifically his collection of fishes and amphibians, was the only one of its kind and that it formed an 'exquisite ornament' for the city of Berlin.[214] They entreated the king to make haste; if he did not procure the collection before it was put to auction, they argued, it might be dismantled at the hands of eager buyers, or worse, remain intact but carried off abroad by one those foreign collectors who had been anticipating the collection being put up for offer. Their attempts to save it from such a glum fate were eventually successful: Friedrich Wilhelm bought it, securing it for a mere 4500 Reichsthalers, half the asking price. Bloch's collection became, officially, a national asset.

---

211 The transfer is documented in "Acta betr. die Übergabe des Blochschen Kabinetts an Prof. Walter. 1804", BBAW, PAW (1700–1811), I-XV-30; f49r–54v contains a floor plan for the room in which Bloch's collection was to be displayed, that shows a designated spot for the chests of drawers in which dried fish could be held. The file also contains a design for these cabinets.

212 Martin Heinrich Klaproth and David Friedländer to the King's Secretary, undated, BBAW, PAW (1700–1811), I-XV-29, f1or.

213 See also: Eva Dolezel, *Der Traum vom Museum: die Kunstkammer im Berliner Schloss um 1800: eine museumsgeschichtliche Verortung* (Berlin: Gebr. Mann, 2019); Marcus Becker, Eva Dolezel, Meike Knittel, Diana Stört and Sarah Wagner (eds.), *The Berlin Kunstkammer: Collection History in Object Biographies from the 16th to the 21st Century* (Petersberg: Michael Imhof Verlag, 2023).

214 *Tagebuch*, MfN, ZMB, GNF S. Bloch 1, TB 6, 67 B 3102 a-g, 108r/v; I have made use of the transcription by Doreen Bombitzki, dated 16 June 1999 in ZMB, S. Bloch 1, bd. 4, 23–24.

Conclusion: Shared Sites of Investigation

Depending on who you asked, in the early modern period an answer to the question 'what is a fish?' may have included a jellyfish, a perch, a flair, a siren, or a whale. Sixteenth-century conceptions of fish were to a large extent synonymous with *aquatilia*: animals that dwelled in the water. Ray and Willughby, however, defined a fish as a creature without feet and with fins that never willingly came onto land. Artedi slightly altered that definition, adding that fishes might occasionally take to the land. Linnaeus, in the tenth edition of the *Systema naturae* (1758), declared whales to be separate from the fishes and included them in his newly created category of mammals. Bloch proclaimed that he would not adhere to this definition, and that he took both whales and swimming amphibians as fishes. In 1804, Bernard Germain de Lacépède (1756–1825) gave the whales their own series rather than discussing them in his books of fish.[1]

Even something as seemingly simple as a fish is a complicated concept. As the introduction to this book has already suggested, discussions about the nature of a fish continued well beyond the eighteenth century. This book, however, has been not so much about these changing definitions themselves, but rather about what such shiftings of meaning might signify. They show, for instance, how theories are formed, what kinds of evidence are deemed fit to test these theories with, and through what sources and methods such evidence can be gained. As this book has shown, the underwater world has always been a shared site of investigation. A wide variety of practitioners, from practical experts to learned savants, busied themselves with the animals that dwelled in ponds, streams, rivers and seas. This book has examined how the field of ichthyology took shape by a dynamic process of construction that entailed continous boundary work. Such work took place not only on the level of the object of study (the definition of what, actually, a fish is), but also with regard to its methods and materials.

Who possessed authoritative knowledge about fish over the course of the eighteenth century, and to what extent did this change over time? This conclusion discusses two central concerns of this study has been the development of methods to examine fish: that of abovementioned methods and materials. I argue that the emphasis of the study of fish changed from the literary to the experiental, to the taxonomic, and argue that this constituted a

1 Bernard Germain de Lacépède, *Histoire naturelle des poissons*, and his *Histoire naturelle des cétacées* (Paris: Plassan, 1804).

form of disciplining. To gain ichthyological knowledge required one to look at these finned creatures in a particular way, and to record and preserve specific properties. This process of disciplining both shaped and was shaped by the different practices outlined in the introduction: classification, description, illustration and collection. These changes will first be explained through methods, and subsequently through materials.

1 The Ichthyologist and the Artisan

In the long eighteenth century, the word 'ichthyologist' came to be used for those who examined fishes through a method of classification. This section looks at that development, and how the knowledge of artisans acted as a foil to what came to be denoted as ichthyological knowledge. First, however, it is important to stress that naturalists who studied fishes in the long eighteenth century seldom did so without taking other aspects of living nature into account. Each of the naturalists discussed took to the study of natural history as a whole. Willughby and Ray examined plants, insects, fishes, birds and quadrupeds; Artedi was inspired by de Tournefort's classification of plants and trees, and used it as a model in devising classification schemes not only for fishes but also for mammals; and Bloch's collection, despite its focus on fishes, covered the whole range of nature, from birds to polished stones, from shells to insects.

They were, furthermore, often engaged in natural philosophical debates, which in the eighteenth century tended to focus on the search for patterns and regularities in order to uncover laws that explained how nature behaved. Such natural philosophical ideas, like Descartes' mechanist philosophies, transformed natural historical study and the understanding of living creatures. Conversely, observations and experiments on fishes (like inserting specimens into an air pump, examining spawn, or investigating the shocks produced by electric eels) could be used to prove or disprove theories on the nature of air, procreation or electricity. The study of nature remained an intertwined endeavour that accommodated a broad curiosity about the world and its creations.

It is within this capacious quest for knowledge that we see a process of specialisation take place. As Martin Rudwick has suggested, in the eighteenth century, learned men and women came to apply themselves to (and establish their reputation in) one or two specific fields, even if their interests ranged rather more widely.[2] This process of specialisation entailed an elaborate discussion

2 Rudwick, *Bursting the Limits of Time*, 48.

on the inclusion – and, consequently, the exclusion – of both the methods and materials that were of import to the study of fish. In this section, I argue that from the sixteenth to the eighteenth century, the methods through which fish were examined underwent an important change, reshaping the perception of who held authoritative knowledge about these creatures. This transformation positioned the ichthyologist at the center, with classification as the defining practice of their expertise.

What makes the study of fish particularly interesting when compared to many other classes of animals is the fact that, throughout the early modern period and in all the corners of the world, fish had a considerable economic value. They were caught and traded in high volumes by fishermen and fishmongers, who as practical experts were not part of the learned class but did have privileged access to the world underwater as a place of investigation. It was their sustained experience, built on years of passing fishes through their own hands, that made their judgements particularly valuable to naturalists. They caught and collected fishes, delineated between species, commented upon the procreation of fishes, and explained how certain species were best caught as well as how they could be cured for trade and consumption. All these matters resurface when perusing the pages of eighteenth-century fish books, but can be drawn out further by combining a study of these books with that of surrounding sources such as correspondences or manuscripts as this work has done.

Chapter 1 sought to explore the significant changes that occurred in the study of natural history between the sixteenth and seventeenth centuries through a discussion of the *Historia piscium*. In the time of humanists like Belon, Rondelet and Gessner, natural history was on the one hand a bookish enterprise. They pored over an extensive body of literature that ranged from the works of ancient naturalists to cookery books, medicinal texts and agricultural treatises. On the other hand, they compared and contrasted what they read in these books with that which they saw around them, taking to the field and examine nature in the flesh. Willughby and Ray, however, envisioned a natural history based solely on observation, rid of ancient writings. The chapter has shown how, in the seventeenth century, the study of fish shifted from the bibliographical, literary method of the humanists to one that foregrounded experiental knowledge that was gathered through the senses (although as, we saw, the written word continued to an important source of knowledge). The result of this strong focus on empirical observation was that the experiences of people of practice, long valued when it came to acquiring natural knowledge, were actively sought out.

In Chapter 2, we saw the evaluation of these experiences took place in practice. It argued that the boundaries between 'theoretical' and 'practical' engagements with fishes could be blurred. Willughby, Ray and their fellow Royal Society members focused on the physical characteristics of species so as to demarcate one from another, examining their properties through empiric observation and the performing of experiments. As it turns out, fishermen and fishmongers deployed similar methods: they distinguished between species based on their physical structures, traced the yearly growth of salmon, for example, and in the case of Baldner, held the claims made in books up to empirical scrutiny. The experience of these individuals, which here has been denoted as practical, artisanal knowledge, was an important part of natural history.

This book has submitted that the underwater world was a shared site of investigation. But it has also examined *how* it was shared, at least from the perspective of the naturalist. As we saw, fishermen, fishmongers and cooks engaged in the study of fishes in distinct ways, even if not all their inquiries were documented in writing and even though few of those contributions that are mentioned in books and letters can be traced back to specific individuals. These written records also have their limits. As interactions in practical and artisanal communities would likely have been of an oral nature, such exchanges are difficult to reconstruct and thus often overlooked in early modern networks of knowledge. In writing up their accounts of these exchanges, naturalists produced edited and condensed reports, performing 'acts of anonymous acknowledgement' that purposefully obscured the names of those who had shared their knowledge. There was thus power in publishing. Those members of practical and artisanal communities with enough financial means to produce manuscripts, such as Baldner and Coenen, were able to circulate their views on nature under their own name.

Chapter 3 has focused on how Artedi propounded classification as a privileged method for knowing fish. The preceding chapters already indicated that classification was an important concern of natural history from the sixteenth century onwards. While this century saw a vivid tradition of sorting species of *aquatilia* into various groups, based on a varied range of ordering principles, Willughby and Ray proposed that they need look only at morphological features in assembling species into groups and delineating one from the other. They focused on characteristic marks, carefully counting and measuring the physical properties of species whenever possible. Ray also put forward an explanation of what a species was, a concept that, despite having been in use for centuries, had seldom been defined. Artedi, however, elevated the practice of classification to the core of natural history. According to him, the goal of natural history was to divide species into neatly devised hierarchical tiers, with the genus as the linchpin. He thus established and defined the taxonomical ranks of classes, order, genera, species and varieties for fishes, just as his friend Linnaeus did for other animals, plants and minerals.

Artedi equated the natural historical study of fishes, which he called ichthyology, with the three-tiered method for classification he devised. As we saw, he defined ichthyology [*ichthyologia*] as "the science [*scientia*] that first specifically indicates all parts of the Fish, subsequently exposes the true Genera and Species names, and finally commemorates the noteworthy Characteristics observed."[3] This triad of indicating and naming parts, naming species and genera, and exposing particularities, became the dictionary definition of ichthyology. Artedi deemed artisanal knowledge of fishes 'amethodic' and thus not of relevance to the study of ichthyology. Artedi's system was based on the careful enumeration of pertinent physical characteristics. His species descriptions consisted of numerical, almost formulaic lists of such characteristics as he sought to make natural history less *historia* and more *scientia*. This attempt fits into the increasingly quantitative bent that, besides natural history, was introduced into other fields of study, such as that of the body or the weather. This epistemological shift emphasised regularity over variety. In order to render variety regular, classification was key.

Classification was thus a form of demarcation that not only separated one species from another, but also separated perceived methodic from a-methodic knowledge. Artedi's classification system strongly emphasised the counting of certain parts of the fish, like teeth or fins, rather than paying attention to other features such as colour or habitat. Selecting those particular features was practical, because matters like colour and habitat could no longer be gleaned from a fish once it had been taken out of its environment and preserved as a specimen in a collection. This classification method also suggested a certain reliability, perhaps even replicability: everyone so inclined could try to find an exemplar of the species under discussion, count the rays in the fin, and ascertain the validity of the reported observation for oneself. In a way, the system that Artedi devised narrowed down the kinds of knowledge that were required for inquiries into fish, implying that true, authoritative understanding no

3 "Ichthyologia est scientia, quae primum omnes Piscium partes nominatim indicat, deinde nomina Generica et Specifica vera ostendit, & denique Proprietates quasdam observatu dignas interdum commemorat." *Icht.*, *Philosophia*, aph. 5, at 2. Original in emphasis.

longer stemmed from the daily identification of species but rather from the ability to construct a classification scheme for God's creatures.

Finally, Chapter 4 moved to a celebrated naturalist who adopted and refined Artedi's system. Bloch, with his impressive collection of fish, became one of the most renowned ichthyologists of his time. In demonstrating his mastery of the method of classification through both his cabinet and his books, he claimed a position for himself in the learned community, no mean feat for a Jewish German. While he was very much invested in the artisanal knowledge of fishermen, fishmongers and cooks, he mostly regarded this kind of knowledge through the lens of oeconomy, as was the fashion of his time. As we saw, both his collection and his fish series benefited from colonial infrastructures, with contributors in South India collecting and preserving specimens, divulging local taxonomies, and creating illustrations. His sumptuously illustrated work raised important questions on how best to capture a fish on paper, and ultimately touched upon fundamental questions of the materials of natural history.

In placing the ichthyologist in opposition to the artisan, we shed light on the changes that had taken place in the study of fish by the end of the eighteenth century. In this juxtaposition, the category of the artisan stands for the wider world of embodied knowledge that fishermen, fishmongers, anglers and cooks possessed. Over the course of the period studied in this book, this kind of practical and artisanal knowledge of fish was gradually pushed out from what became 'formal fish knowledge', i.e. the kind of knowledge about fish that was codified in natural historical books. This was, also, only a theoretical opposition. As we saw, the knowledge of communities whose experiences of nature stemmed from practice and craftmanship continued to be incorporated into such works – it was simply no longer stressed as authoritative knowledge. Throughout its chapters, therefore, this book has argued for the interrogating of concepts such as 'practical' or 'local' to ascertain what these really entailed, rather than using them to obscure relations and identities.

Natural history, as conceptualised by learned naturalists, was not an activity that was inherently separate from other activities revolving around the natural world, such as hunting, fowling, fishing, livestock breeding or herbalising. This book has argued that the fact that these were, eventually, separated was the outcome of a conscious process of demarcation. A comparative study of other natural historical pursuits, such as the branches that would be given the names of geology, paleontology, herpetology, ornithology or entomology could further elucidate the extent to which the ways in which these boundaries were drawn up was particular to the study of fish. Barnett, for example, has stressed the parallel processes of showing and hiding of labouring-class fossil suppliers in the field of earth science, whereas Hünniger has similarly drawn attention to the social intracies of bug collecting in the field of entomology.[4] Situating these exchanges more firmly within the dynamics of discipline formation could serve to better understand how practical or artisanal knowledge, and in particular the holders of that knowledge, were excised from natural history over the course of the long eighteenth century.

The disciplination of ichthyology took place through the practice of classification. This was not so much because those from practical, artisanal communities did not distinguish between fishes – we have seen that they did – but because it elevated the ichthyologist into a specific, methodic authority on fish. This separation (re)constructed this mundane, everyday object into an object of scientific study. Underlying these kinds of discussions on method were pressing epistemological questions with regard to the materials of natural history. After all, natural history revolved around a physical, at times messy, world that nevertheless demanded interpretation. The following section explores how the meaning of these materials for the study of fishes changed over the course of the eighteenth century. It focuses on the connections beween practices of collecting, describing and illustrating, and how these both shaped and were shaped by this disciplination of natural history.

2 Fish as Specimen, Text and Image

A second concern of this book is the matter of representation of fish, either in the form of a preserved specimen, in word or in image. Fish, after all, were thoroughly unstable objects of inquiry. The chapters in this book have given examples of the myriad ways in which this held true. Glutinous matter might eminate from a fish prepared as a wet specimen, requiring a refreshing of the alcohol in which it is immersed. The fin of a dried specimen – crucial for the classification of a species – might easily rub off. A specimen's scaly skin might crack. No amount of varnish, however well applied, could preserve a fish's colour. In time, colours fade or mellow, turning to brown hues. Fish specimens might arrive in the cabinets of collectors altogether spoiled. Preserved specimens thus posed their challenges, as did illustrations and textual descriptions. These uncertainties of representation were one of the reasons why first-hand experience of living fish was particularly prized.

4 Barnett, "Showing and Hiding," 245–274; Hünniger, "Visible Labour?" 180–210.

Epistemological questions of how to represent fish were heatedly discussed by naturalists, artists and collectors alike. This section traces how these debates changed over the course of the eighteenth century, and shows how such tensions were heightened by the increasing value that was attached to classification of fish through specific physical features. It argues that the formulation and circulation of three-tiered classification principles trained naturalists to look at fish in a particular way, and that this impacted the way in which they produced their specimens, descriptions and illustrations (or, how they had these produced by others). Although this book has stressed the extent through which these practices were intertwined, for the sake of clarity it first looks at collection, then description, and, finally, illustration.

The practice of collecting serves as a good entry into this discussion, as it confronted naturalists with the dazzling diversity of nature around them in a very physical, concrete sense. By the term 'collecting', I mean the extraction of species from the water, the subsequent act of preserving specimens, and the creation of a cabinet through the assembling of specimens into one place. In the cabinets of curiosities of the sixteenth century, both human-made artifacts and natural specimens were placed alongside each other, allowing the onlooker to meditate on their affinities or differences to gain a better understanding of the material world; think, for example, of Felix Platter's cabinet. Over the course of the seventeenth and especially the eighteenth century, such encyclopeadic collections gave way to repositories that focused on natural historical objects, and even specific classes of animals. While Paula Findlen has argued that collections passed up on breadth as naturalists 'stopped seeing', one might say that they developed a different way of looking.[5] An animal like the sea horse was no longer displayed in a collection for the fascinating structure of its skin or for its interesting shape, but rather as a specimen that, on account of certain defining marks, should be included to complete a series of similar species.

The classificatory gaze thus directed both what was collected and how these items were organised into collections. The natural specimens in them were neatly arranged according to contemporary classification schemes, be it Martini's method for categorising shells or Linnaeus' method for fish (which was largely based on that of Artedi). By displaying their objects in this way, naturalists both displayed their knowledge of existing classification systems and showed they were serious naturalists. Towards the end of the eighteenth century, the number of known species had grown to such an extent that it was

5 Findlen, *Possessing Nature*, 405.

no longer considered possible for naturalists to strive for a complete collection of all classes of plants and animals.[6] They would therefore opt to place their focus on a specific class. These collections would be closely tied to their reputation as authorities on that specialised group, as we saw with Bloch.

The fish in these collections would, by and large, be preserved in either of two ways: as dried or as wet specimens. The first option was laborious, the other expensive. The drying of fish in such a way that they would not smell required considerable handiwork, carefully cutting fish open, removing their flesh and stuffing the skins back into shape with hay. Immersing a specimen into alcohol entailed less work, but the materials were costly. In the eighteenth century, Gronovius debuted his method for preserving the skin of fish into print. He advertised it as both inexpensive and easy. Those characteristics most salient for classification, such as the fins and their rays and the bones in the gill flap, could be pressed into the service of the inquisitive naturalist. The preservation method facilitated the circulation of fish skins, flattened enough to fit into a letter wrapper, across the various corners of the world. As present-day efforts to replicate this method have made clear, however, it required no little dexterity in wielding a scalpel and some ability to read between the lines of instructions.[7]

To what extent could a description stand in for a specimen? As Brian Ogilvie has argued, the sixteenth century saw a standardisation of how to describe living nature. As the study of particulars, after all, natural history lent itself well to verbose descriptions. In the sixteenth century, the humanistic, literary way of describing living nature rich with allegorical reference transitioned to a manner of description that emphasised the physical characteristics and marks that distinguished one species from another. Willughby and Ray emphasised direct observation in their descriptions. They stressed the ideal of *autopsia*, or seeing with one's own eyes, and when possible perfomed dissections on specimens – literally cutting to the bone. A good proportion of their descriptions,

6 Dominik Hünniger, "'Extolled by Foreigners': William Hunter's Collection and the Development of Science and Medicine in Eighteenth-Century Europe," in *William Hunter and the Anatomy of the Modern Museum*, eds. Mungo Campbell, Nathan Flis and María Dolores Sánchez-Jáuregui (New Haven: Yale University Press, 2018), 135.

7 See: Barlow Robles, *Curious Species*, 172–183; Barlow Robles, "Natural History in Two Dimensions," *Common Place: The Journal of Early American Life* 18, no. 1 (2018); Didi van Trijp and Robbert Striekwold, "The Ichthyologist's Garden," The Recipes Project (2017), https://recipes.hypotheses.org/9798 (accessed February 5, 2025) and Luis Ceríaco and Mariana Marques, "Peixes Em 'Herbário.' Uma Técnica Científico-Museológica do Século XVIII," Conference Paper, *Congresso Luso-Brasileiro de História das Ciências* (October 2011): 1204–1219.

therefore, provided elaborate anatomic detail of fish in a way that preserved objects or illustrations could not.

Artedi introduced a format for the description of species in the form of a numerical list. This list contained those characteristics that he found salient for identification and classification: *viz.*, the number, shape and position of fins, the number of rays in the fin, the number or shape of the teeth, or the number and shape of other parts. This quantative approach lent itself well for expression as a formula which contained a letter referencing the fin or body part and a number representing the quantity thereof. Because it was not only the number of the fin rays, but also their quality (thorny or soft) that counted, the division between these types of fin rays was sometimes conveyed as a fraction. The species of the common scad, then, could be condensed as follows: "P. xx. V. vi. A. ii/xxi. C. xx. D. viii. xxxiv.", referring to the pectoral fin (P), ventral fin (V) anal fin (A), caudal fin (C) and dorsal fin (D) respectively, the Roman numerals indicating the amount of fin rays.[8] Even although Bloch's descriptions could fan out for pages, expanding, for example, on the oeconomic meaning of a particular fish, each and every species description opened with this kind of 'diagnostic' formula.

In theory, Artedi's numerical list would make illustrations superfluous. Here, we are reminded of Gronovius' statement that all he cared to know was the number of bones in the gill flap of a specimen, so that he could determine its genus as expressed within Artedi's system. If he could ascertain this number based on a preserved specimen, he had no need for an image. In this, he pitted himself against naturalists such as Catesby, who included lavishly executed illustrations of fish along with their descriptions. Gronovius admitted that these images were beautifully done, but found that they ultimately served no classificatory purpose. Here, Gronovius made an epistemic choice: objects were preferable to images (which could be wrong) as a basis of classification. What then, was the use of images?

Illustrations continued to be crucial to the natural historical study of fish throughout the early modern period as an important way of capturing fishes on paper. One of the reasons was their power of preservation. As fish lost their colours soon after they had been lifted from the water, it was preferred that a fish be depicted while as fresh as possible, so that the qualifier 'from the life' took on a quite literal meaning. Images made the colour of a species endure beyond its natural decay and could also render its shape intact, the fins fanning out as if it were still under water. The body shape and the form of fins,

8 Bloch, *Allg. Nat. der Fische*, vol. 2, 104. Currently accepted species name: *Trachurus trachurus*.

after all, were similarly difficult to retain in dried specimens, which tended to shrink and wither. As we saw, artists deployed several techniques for locking the colour of a specimen in their drawings. Catesby, for example, had a succession of freshly caught fishes painted, whereas Bauer added numbers that corresponded to a specific colour scheme to his fish sketches, adding these colours to his final drawing a later stage. A particularly noteworthy technique was the use of paints containing silver and gold pigments. As the metals in the paint reflected light, artists could play with its luminescent properties. Applying the paint to areas of the fish that naturally catch light, like the edges of its scales and gill cover, they gave the illustration a lively quality.

This is not to say that illustrations acted as proxies for specimens. In fact, the various chapters have thrown light onto the many epistemological decisions that the making of images entailed. Naturalists might have different aims in mind for their illustrations. Images might act as supporting evidence for the existence of a particularly strange specimen like the mermaid, as tools to communicate knowledge about an animal with ease and pleasure, a mnemonic device for the identification of a particular fish, or as representations of a specific species that could serve as the basis for classification. Such aims, of course, did not necessarily exclude one another. However, in creating these 'epistemic images' naturalists left out certain features and highlighted others. The artists and colourists with whom these naturalists worked also took part in decisions for the execution of images; preferably by mutual agreement, though this could go awry. Practical considerations played a role in how images turned out. Image-making was a time-consuming and costly enterprise, and could lead to no little frustration.

Throughout the early modern period, fishes were depicted according to strikingly stable pictorial conventions: portrayed from the side against a white background. With some exceptions, like Baldner's manuscript which depicts its aquatic creatures in small pools of water, illustrations by and large adhered to 'specimen logic', in which a species is visually extracted from its environment and presented as a decontextualised specimen. The remarkable changes in the execution of images occurred more on the level of technique. During the period examined in this book, several developments created new possibilities for illustrative practice. Chapter 1, for example, has discussed how artists could, compared to the possibilities presented by wood block technology, work ever finer detail into copper engravings, and Chapter 4 showed how artists made use of new inventions in colour printing to intensify the hues of their natural historical illustrations.

Illustrations drawn 'from the life' continued to carry special weight, from Belon to Bloch. The visual strategies with which artists conveyed that a drawing had been made from the life, however, differ markedly. Certain illustrations in Willughby and Ray's *Historia piscium* were endowed with a 'rhetoric of the real', in the words of Kemp, by having depicted objects cast a shadow on the page. Bloch's inclusion of colour served to suggest that the illustrations had been done after living specimens (even if this was not always the case). The perishable nature of fish made claims of direct observation crucial to authority, and this is evident in the visual rhetorical strategies that suggested a proximity to the object at hand in the illustrations that trumpeted such claims.

The classificatory sway of the eighteenth century is also visible in the illustrations that were produced during this period. A clear example of this is the instruction that Bloch gave to the artists in his studio tasked with capturing the fish in his collection on paper: namely, to depict precisely the number of fin rays so that the specimen could be classified based on the engraving in the book. That he was committed to seeing this policy carried out is clear from the extant drawings that served as the models for these engravings, where the number of fin rays were noted down for each of the depicted fins. As Bloch deemed colour relevant for the purpose of classification (even if Artedi rejected it as too unreliable a feature to be used in this way), he required his colourists portray the natural colour of the fish. As his book series demonstrates, a luxuriously executed hand-coloured engraving, accentuated with paint of real silver or gold, did need not to be antithetical to the practice of classification.

Uncertainties regarding the level of trust that one could place in a specimen, in a description or in an image were prevalent throughout the early modern period. In the eighteenth century, however, such anxieties increasingly focused on the ways in which specimens, texts and images encapsulated those features that were pertinent for classification. How to be sure that the specimen you received was complete? How to trust that an artist had portrayed the right amount of fin rays in his or her drawings? Was it judicious to depend on the description of a naturalist, who might have based his rendering on an inaccurate image, or may well have used a different method for counting the fin rays? Such problems were not easily solved, but the debates offer fascinating insight into how naturalists evaluated the respective reliabilities of certain representations of a fish over other ones.

3 Disciplined

The development of natural knowledge is a far from straightforward, linear process. In the study of fish, early modern naturalists build on each other's works while also taking distinctly new directions, or returning to definitions, methods and practices that had been cast aside by those who came before them. For example, where fishermen and fishmongers were vital sources of knowledge for Willughby and Ray and credited as such in their publications (albeit in anonymous acknowledgement), Artedi only mentioned them in his work as a category of people who dealt with fish whose distinctions were not relevant to what he defined as ichthyology. Artedi did not include images and his very method perhaps made them superfluous, yet Bloch decided to only discuss those fish species in his work of which he was able to offer illustrations that had been done after life as he deemed images essential in representing what a fish was.

As Kelley has stated, the study of disciplines has tended to be partial to intellectual continuity and ideas of progress, leading to a presentist and internalist approach.[9] If we assume fish to be a coherent category, and we take the concept of ichthyology in its current form for granted, we miss the vibrant culture of the long eighteenth century in which observations and ideas about the peculiarities of living nature were exchanged abundantly. With its attention to processes of boundary work and acts of demarcation, as laid out in this conclusion, this book hopes to present a way to study similar developments in other fields of knowledge. To be attuned to the human efforts involved in the demarcation of disciplines, after all, is to understand that the process of the separation and specialisation of knowledge that we can discern both in and after the eighteenth century is a profoundly historical one. The arts and the sciences, practice and theory: all were used to form a comprehensive ideal through which we might make sense of the world.[10]

In 1828, Georges Cuvier published his *Tableau historique des progrès de l'ichtyologie depuis son origine jusqu'à nos jours* [Historical overview of the progress of ichthyology from its origins to the present day]. This generalist, whose expertise ranged from fossils to comparative anatomy to fish, had decided to draw up his own scheme for the classification of fish. This historical overview formed the first part of what would become the 22-volume *Histoire naturelles des poissons*. He contended that such a recapitulation was necessary in order to understand how this branch of natural history had developed.[11] To be able to consult earlier authors who wrote on fish in a worthwhile way required, as he

9 Kelley, "Introduction," 3.

10 See also: Wijnand Mijnhardt, "'The World We Have Lost': In Praise of a Comprehensive Concept of Science and Scholarship," in *Teyler's Foundation in Haarlem and Its 'Book and Art Room' of 1779*, eds. Ellinoor Bergvelt and Debora J. Meijers, esp. 84–86.

11 Carolyn Scearce, *Fish Facts: Disciplinary Development of Ichthyology in Nineteenth-Century Europe* (PhD diss., University of Oklahoma, 2019), 31.

stated in historicising fashion: "[...] to know the circumstances that governed their work, the times in which they lived, the condition in which they found the science, and the facilities procured for them either through their personal position or through the help of friends, patrons, or students."[12] He wrote the first history of ichthyology, a field that by the early nineteenth century had acquired enough of a distinct shape on the tree of natural history that it merited its own historical narrative. The approach he unfolded was a linear one: he saw ichthyology going forward "unobstructed towards perfection", with his book series as the culmination of all these previous efforts.

This continued historical process of disciplinary division is also evident in contemporary museum collections. Early modern encyclopaedic collections in the homes of physicians, apothecaries, merchants and professors were, over the course of the eighteenth and nineteenth centuries, relegated to newly founded public collections, often with national and/or royal status. Their various contents were organised according to the concept of disciplinary categories that was reigning at the time either upon arrival, or during a later moment when they were transferred to other heritage institutions. A good example is Hans Sloane's wonderfully rich collection of papers, artifacts, and specimens which comprised the founding collection of the British Museum on its institution in 1753, and of which parts were, in the twentieth century, transferred to the British Library (papers) and to the Natural History Museum (specimens) – although the skin of a scorpion fish pasted on paper, straddling the divide of paper and specimen, still swims amongst the shelves of the storages of the British Library.[13] Other European examples of public museums that absorbed private collections are the *Muséum national d'histoire naturelle* (1793) in Paris and *'s Rijks Museum van Natuurlijke Historie* in Leiden, founded in 1820.[14]

Bloch's collection underwent a similar fate. In 1802, the entire contents of his cabinet were relocated from his home at the Spandauerstraße to King Friedrich Wilhelm III's *Kunstkammer* in the Berlin Palace. As natural history had not been a strength of his collection, the king sought fit to commission the purchase of private natural history collections, for which a new space was established.[15] Bloch's fish collection was designated its own room to facilitate

---

12 Georges Cuvier, *Historical Portrait of the Progress of Ichthyology form Its Origins to Our Own Time*, ed. Theodore W. Pietsch, trans. Abby J. Simpson (Baltimore: The Johns Hopkins University Press, 1995), 21.

13 Delbourgo, *Collecting the World*, XXV.

14 On the latter, see: Maria Eulàlia Gassó Miracle, *Coenraad Jacob Temminck and the Emergence of Systematics (1800–1850)* (Leiden: Brill, 2022), 75–89.

15 Meike Knittel, "Eurasian Golden Plover Zmb Aves 13021: An Original From The *Kunstkammer*?," in Becker et al., *The Berlin Kunstkammer*, 180.

research and teaching.[16] Eight years later, the natural historical specimens of Bloch's collection were again packed into crates. While his paintings stayed in the *Kunstkammer*, his specimens were transported to the newly opened Universität zu Berlin to form part of its zoological museum, established by the King at the instigation of Wilhelm von Humboldt.[17] After a few decades in the research and teaching collection at the university of Berlin, Bloch's natural historical specimens moved house once more. They were transferred to the *Museum für Naturkunde* in Berlin, where they remain today. Since then, all of the objects in his collection have been divided according to the class to which they belong: Bloch's fishes are part of the fish collection, his reptiles reside among the collection of reptiles and amphibians, and so on. His capacious collection has thus been separated into distinct categories, sorted into specialised departments, awaiting what the future might have in store for it.

16 Eva Dolezel, Meike Knittel, and Diana Stört, "Around 1800: The *Kunstkammer* in Transition," in Becker et al., *The Berlin Kunstkammer*, 175.

17 Peter Giere, Peter Bartsch, and Christiane Quaisser, "From Humboldt to HVac – The Zoological Collections of the Museum für Naturkunde Leibniz Institute for Evolution and Biodiversity Science in Berlin," in *Zoological Collections of Germany: The Animal Kingdom in its Amazing Plenty in Museums and Universities*, ed. Lothar A. Beck (Dordrecht: Springer, 2018), 95.

Epilogue: Using Historical Sources to Understand Ecologies of the Past

It can be all too easy to forget, but the natural world inhabited by the fishermen, fishmongers, divers, traders and missionaries discussed in this book was different from ours today. Species that might have been common in the eighteenth century might now be rare or, as in the case of the European eel or angel shark, even critically endangered.[1] This realisation necessitates a reflection on the role of the historical sources of natural history in understanding past occurrences and distribution which can be of help in protecting these species and their aquatic environments now and in the future. This epilogue set outs some of the opportunities, challenges and considerations in using historical sources in this way.

Such a reflection can be guided by the notion of baselines: reference points for the occurrence of a species. These standards lie at the basis of policies for contemporary conservation efforts. The problem with baselines, however, is that they shift. Interestingly, this problem was first articulated in the context of fish, when fisheries scientist Daniel Pauly introduced the concept of the shifting baseline syndrome in a short but influential article. He explained it thus: scientists accept as baseline the stock size and species composition that they encounter at the beginning of their careers, and use this to evaluate change. When the next generation starts its career, the stocks have further declined, but it is the stocks at that time that serve as a new baseline.[2] Reference points for the occurrence of species thus change imperceptibly yet radically. This also extends to the public at large, who might find it hard to believe, for example, that the grey whale could commonly be spotted from the coasts of the North Sea during the seventeenth century.[3] The largest issue is that nature conservation agencies set measures in order to return to a certain baseline, unaware that this very baseline itself is subject to change. As a result of this, the policies

1 Alec Moore and Jan G. Hiddink, "Identifying Critical Habitat with Archives: 275-Year-Old Naturalist's Notes Provide High-Resolution Spatial Evidence of Long-Term Core Habitat for a Critically Endangered Shark," in *Biological Conservation* 272 (2022): 1; Cinzia Podda, Francesco Palmas, Antonio Pusceddu and Andrea Sabatini, "When the Eel Meets Dams: Larger Dams' Long-Term Impacts on *Anguilla anguilla* (L., 1758)," in *Frontiers in Environmental Science* 10 (2022): 8.

2 Daniel Pauly, "Anecdotes and the Shifting Baseline Syndrome of Fisheries," in *Trends in Ecology and Evolution* 10 (1995): 430.

3 Marc Argeloo, *Natuuramnesie: Hoe we vergeten zijn hoe de natuur er vroeger uitzag* (Amsterdam: Atlas Contact, 2022), 171–172.

they draw up for the conservation and preservation of species – despite coming from the best of intentions – are far less ambitious than they could be.

Solution is to be found in anecdotes, according to Pauly. By this he means testimonies of past large catches that bring historical consciousness to a field of research that is very much focused on the present. These testimonies can take the form of oral history, for example by interviewing multiple generations of fishermen.[4] Fisheries scientists, fish biologists and fish ecologists can also profit from the natural historical sources discussed in the present book, as these offer insight into the presence of certain species in previous centuries, and indicate whether these, for example, were once common stock in regions from which they have since disappeared. This is where the tool kit of the historian is particularly handy. Especially when it comes to sources that stem from the early modern period or earlier, several kinds of expertise are required. These include archival know-how, paleographical proficiency, the ability to read neo-Latin, an understanding of the people of places of the period at hand, and a good grasp on historical taxonomy.[5]

Even though historical sources are treasure troves of observations, some caution is needed. Ecologists tend to seek unambiguity: they combine binomial, scientific names of species (and even subspecies) with clear geographical signifiers, preferably with coordinates. Such a level of exactitude is rarely found in historical sources from the early modern period. It can, for example, be hard to discriminate between similar-looking species depicted in manuscripts, books and on paintings.[6] In descriptions, species might be denoted on a more generic level (such as "cetacean" or "species of carp"), whereas the location where they are observed is not always made spatially explicit (like "in our lakes"). What is more, boundaries of territories have shifted over time and these territories might now go by other names.

Despite these challenges, however, there are good examples of how historical evidence can prove useful for revealing past distribution of species populations. An analysis of Adriaen Coenen's *Visboeck* by fisheries researchers yielded general insight into the presence of species in Dutch coastal waters,

4 Andrea Saénz–Arroyo, Callum M. Roberts, Jorge Torre and Micheline Carinõ-Olvera, "Using Fishers' Anecdotes, Naturalists' Observations and Grey Literature to Reassess Marine Species at Risk: The Case of the Gulf Grouper in the Gulf of California, Mexico," *Fish and Fisheries* 6, no. 2 (2005): 121–133.

5 See also: Victoria Pickering, "Mobilising Historical Botanical Data as Research," *Nuncius* 39, no. 3 (2024): 768.

6 Anne M. Overduin-de Vries and Paul J. Smith, "Fishing in the Past: Biodiversity, Art History, and Citizen Science – Preliminary Results," in Smith and Egmond, *Ichthyology in Context*, 311.

which they conclude have now lost part of their richness.[7] More specific information is found in an annotated copy of the *Historia piscium* in the possession of the Welsh customs officer Lewis Morris (1701–1765), to which he added detailed descriptions of his encounters with angel shark (Figure 24). Through an examination of this copy, fisheries researchers have found evidence for the long-term importance of certain stretches of the Welsh coast for this shark, and have argued that these should be appointed a designated area for the conservation of this now critically endangered species.[8]

Historical sources that offer semi-quantitative indications, such as 'rare', 'abundant', 'in great numbers' or 'very well provided' are of particular importance to ecologists. These statements are often combined and contrasted with other kinds of evidence. For example, Hans Sloane's mention of the "great, long prickled Sea Egg", which he found "[...] in great numbers on the reef by Gun-Key, or, Cayos off the Port Royal Harbour" is taken as one piece of evidence for the long-term presence of *Diadema*, a genus of sea urchins in the Caribbean, alongside fossil finds.[9] Other archival sources that give an (indirect) insight into the relative abundance or rarity species include historical economic statistics of the market price of a fish, charter books that detail the regulation of fishing rights, or decrees regarding the trade of fish. In the seventeenth century Dutch Republic, for example, the price of a salmon equaled the weekly wage of a labourer.[10] Taken together, these sources can serve to extend points of reference further into the past, but their interpretation requires care and rigour.

How to reconcile the names for species in various languages? After all, species names come in more languages than the scholar's Latin. As we saw, Willughby and Ray were already well aware that understanding of the dispersal of species constituted a linguistic puzzle. In an attempt to solve this problem, Artedi used no fewer than 109 pages for his *Synonymia nominum piscium* [Synonyms of fish names], a compilation of such synonyms in various languages.[11] A contemporary iteration of this kind of synthesising effort is the online database FishBase, of which aforementioned Pauly was one of the

7 Floris P. Bennema and Adriaan D. Rijnsdorp, "Fish Abundance, Fisheries, Fish Trade and Consumption in Sixteenth-Century Netherlands as Described by Adriaen Coenen," *Fisheries Research* 161 (2015): 384–399.

8 Moore and Hiddink, "Identifying Critical Habitat with Archives," 3.

9 J.B.C. Jackson, "Reefs since Columbus," *Coral Reefs* 16 (1997): 26.

10 Rob Lenders, "The Historical Truth behind the 'Salmon-Servant' Myth," in Smith and Egmond, *Ichthyology in Context*, 460.

11 Further discussed in Aili and Pietsch, *Peter Artedi: Reformer of 18th Century Zoology*, vol. 1, 70–72.

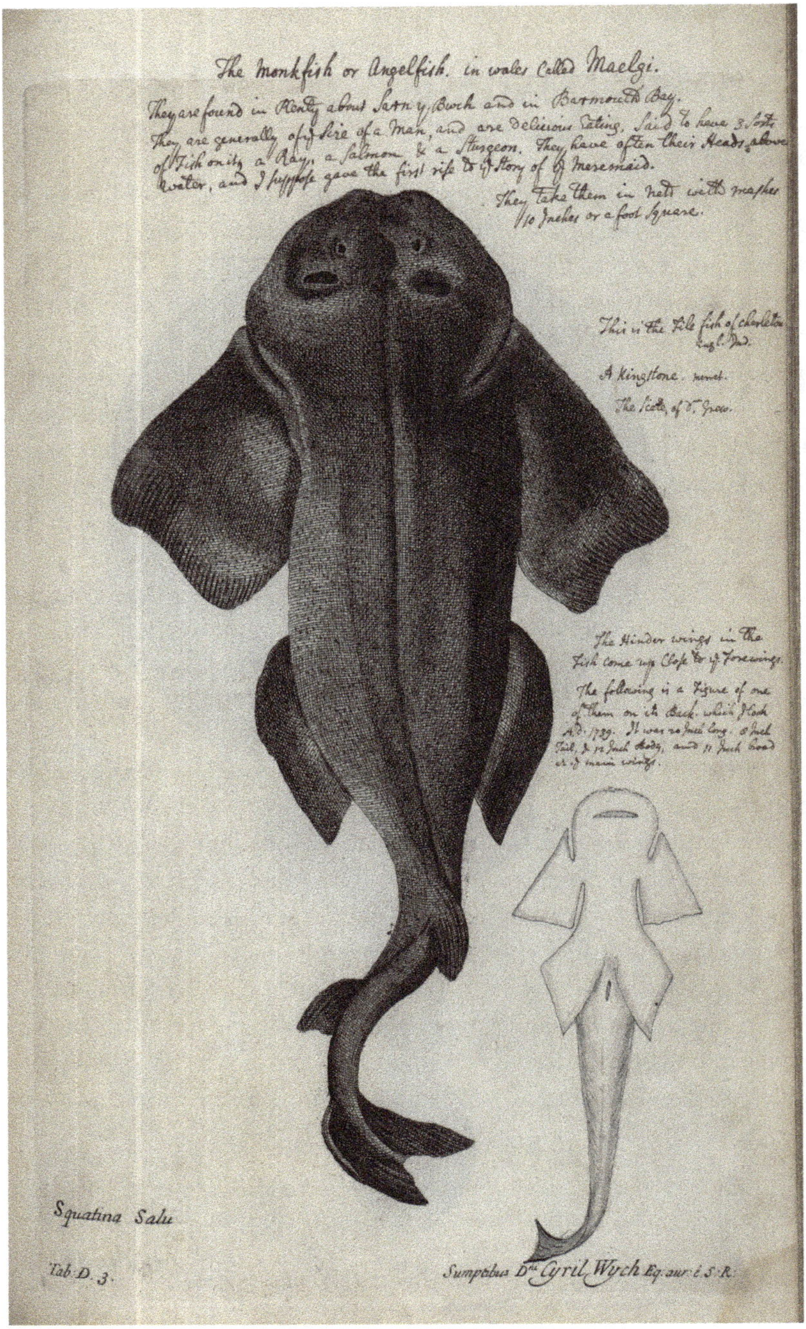

FIGURE 24 Engraving of an angel shark annotated by Lewis Moore. Francis Willughby and John Ray, *Historia piscium* (Oxford: Sheldonian Theatre, 1686), tab D3. Supplied by Llyfrgell Genedlaethol Cymru / National Library of Wales

initiators.[12] This directory incorporates earlier databases, such as Eschmeyer's Catalogue of Fishes which documents revisions of Latin fish names from the tenth edition of the *Systema naturae* (1758) until the present; as this was the first edition of this work to consistently apply binomial species names, it is regarded as the starting point for scientific nomenclature.[13] Other sources on which FishBase draws include recent scientific papers as well as historical publications like that of Bloch and Bleeker. The database includes common names for fish from 350 languages. Efforts such as these help to plot the occurrence of species beyond linguistic borders.

Collating names in various languages, however, is not a matter of assembling interchangeable data points. To the contrary: as the scientists behind FishBase state, fish names offer ecological and biological information as well as important clues to their cultural context.[14] In this regard, Bloch's book series is not only interesting for the Latin names it includes. As Bloch consistently included the local names for those fish that he received from the Coast of Coromandel, his book series also stores valuable information about common names in Tamil. It appears that these historical, common names for fish have not yet been added to databases of fish names. If the many species names that Tamil collectors and other contributors in South India shared with John (who then passed them on to Bloch) were to be added, they could fulfil several research purposes. Because they are connected with the Latin species names offered by Bloch, they expand the list of known synonyms. They could also be etymologically studied, allowing for a historical comparison between early modern and contemporary names of fish, which might yield interesting clues to their spread and usage over time. Last but not least, they point to the colonial context in which these species were collected.

This brings us to an important concern for historians when it comes to the large scale digitisation of natural historical data: namely, the risk of reduction. In the wish to compile data from historical sources, observations of species tend to be reduced to the locality where the species at hand is found combined with its currently accepted species name. This decontextualisation holds for the written sources that have been so far discussed in this epilogue, but mostly for natural historical specimens. Over the past years, millions of the

12 https://www.fishbase.org/ (last accessed 15 February 2025).

13 https://researcharchive.calacademy.org/research/ichthyology/catalog/fishcatmain.asp (last accessed 15 February 2025).

14 Katarina Zimmer, "What's in a Fish's Name?," *Nautilus* (12 March 2024), https://nautil.us/whats-in-a-fishs-name-530331/?trk=feed_main-feed-card_feed-article-content (last accessed 15 February 2025).

natural historical specimens that sit patiently on the shelves of museums have been digitised, usually as part of an effort to share knowledge about species within global infrastructures. With increasing digitisation, however, comes an increased need to think critically about such infrastructures and the kinds of information that they contain.[15]

Researchers have to contend with an absence of data especially when it comes to consulting digitised natural historical objects that stem from a colonial context. Andreas Weber and Esther Turnhout argue that natural historical specimens can and should be kept connected within the context of their colonial provenance by linking the digitised specimens to the rich stores of archival material surrounding them.[16] Far from simply satisfying the curiosity of the historian, this contextualisation builds and sustains a critical awareness of the colonial structures upon which our present-day knowledge of the world's natural riches rests. Furthermore, if this context were to be removed from digitised specimens, past practices of effacing the many colonial contributors of natural knowledge (voluntary or not, as discussed in Chapter 4) would be perpetuated.

There is one further dimension to discuss. Historical sources show us how awareness that the supply of fishes is vulnerable and requires protection is in itself not new. A fascinating example of this is the petition by the guild of London fishmongers mentioned in Chapter 2. In this petition, the fishmongers called on the city parliament to take action in protecting the supply of fish in the rivers, streams and the sea coasts. "The citty of London," they wrote, "hath not bene for some years last past furnished with soe large a supply of fish as formerly it hath bene whereby all sorts of fish doe come to the petitioners' hands at such high rates."[17] A number of causes was listed for the decline of the fish population, alongside measures for improvement. The fishmongers requested to ban the use of casting nets with which the fry and brood are "torne to peeces, & not suffered to come to its maturity", claiming that such a net once killed 1600 salmon smelts in one cast. But they also flagged issues of pollution in areas where dyers and tanners were dumping wastewaters into the river. A further issue that they mentioned are closed flood gates which hindered free

15 Tiziana N. Beltrame, Elena Canadelli, Luca Tonetti, "The Natures of Digital Practices: People, Objects, and Data Mobilities in Natural History Collections," *Nuncius* 39, no. 3 (2024): 741–758.

16 Andreas Weber and Esther Turnhout, "A Langur from Sumatra: Digital Futures, Material Presents and Colonial Pasts," *Nuncius* 39, no. 3 (2024): 787.

17 Petition, RS Cl.P.15i/8, Classified Papers, f1r.

passage for salmon smelts during spawning. The petition appears surprisingly topical.

In sum, historical sources on fish accommodate a wealth of insight. Their value lies on different levels. We can consider them as treasure troves of historical data of species occurrences for those wishing to reconstruct shifting baselines – with all the interpretational labour such data require. They also provide important insight into the global, colonial context in which our contemporary knowledge of the underwater world took shape. Moreover, they are sources for comprehending how people in the past thought about how to conserve the life in rivers, streams and seas. Ultimately, they show the importance of people coming from positions of practice and theory, from the humanities and the sciences to work together to fathom the sheer abundance and variety of species that dwell in on the earth's vast recesses of water, now and in the past.

Bibliography

1 Primary Sources

1.1 *Archives, Libraries and Collections*

1.1.1 United Kingdom

British Library (BL), *London*

Western Language Manuscript Collections

Add MS 3870. Peter Artedi, *Catalogus Piscium Maris Balthici ut et Fluviorum ac Lacuum Sveciae.*

Add MS 5267. Hans Sloane, papers and correspondence.

Add MS 5308c. Anonymous, *A Book of Fishes done at Hamburgh, with Mr Ray's Notes.*

Add MS 6485. Leonhard Baldner, *Vogel-, Fisch- und Thierbuch.*

Add MS 6486. Frederick Slare, translation of Baldner's *Vogel-, Fisch- und Thierbuch.*

Sloane MS 4020. Jacob Theodor Klein, *Petri Artedi operum brevis recensio.*

Printed Books before 1851

956c16. Joseph Banks' annotated copy of Peter Artedi, *Ichthyologia* (Leiden, 1738).

Linnean Society (LS), *London*

Linnaean Annotated Library

BL.144. Carl Linnaeus' annotated copy of Peter Artedi, *Ichthyologia* (Leiden, 1738).

Natural History Museum (NHM), *London*

Gronovius Fish Collection

BMNH 1853.11.12.120. Specimen of *Gonorkynchus greyi.*

Nottingham University Library (NUL), *Nottingham*

Middleton Collection

NUL Mi LM 24/170. Print proof of engraved title page *Icthyographia* (1685).

NUL Mi LM 25/14. Drawing of dissected flair.

NUL Mi LM 25/21. Drawing of pufferfish.

NUL Mi LM 25/51. Watercolour of carp.

Royal Society (RS), *London*

Classified Papers

Cl.P.15i/8. Petition of fishmongers to the parliament, undated.

Journal Book Original

JBO/1/66. Entry of 18 June 1662.

JBO/1/66. Entry of 25 June 1662.

JBO/1/159. Entry of 15 April 1663.

JBO/1/194. Entry of 24 June 1663.

JBO/2/23. Entry of 30 December 1663.

RCN 18574. Annotated copy of Francis Willughby, *Historia piscium* (Oxford, 1686).

National Library of Wales (NLW), *Aberystwyth*

Archives and Manuscripts

MS 24052E. Annotated copy of Francis Willughby, *Historia piscium* (Oxford, 1686).

1.1.2 Germany

Berlin-Brandenburgische Akademie der Wissenschaften (BBAW), *Berlin*

Preußische Akademie der Wissenschaften, PAW 1700–1811

I-XII-11. Circular soliciting plate subscriptions for Marcus Élieser Bloch, *Allgemeine Naturgeschichte der Fische* (Berlin, 1782–1795).

I-XV-29. Acta betr. die Übergabe des Blochschen Kabinetts an Prof. Walter, 1804.

I-XV-30. Letter from Martin Heinrich Klaproth and David Friedländer to the King's Secretary, undated.

Archiv der Franckeschen Stiftungen (*AFSt*), *Halle an der Saale*

Missionsarchiv mit der Indien- und der Amerikaabteilung

AFSt/M 1 C 28 : 87. Letter from Christoph Samuel John to Johann Ludwig Schulze, 18 October 1787.

AFSt/M 1 C 30a : 2. Letter from Christoph Samuel John to Johann Ludwig Schulze, 12 October 1789.

AFSt/M 1 C 30c : 24. Letter from Christoph Samuel John to Johann Ludwig Schulze, 20 January 1789.

AFSt/M 1 C 31a : 21. Letter from Christoph Samuel John to Johann Ludwig Schulze, 20 January 1790.

AFSt/M 1 C 31b : 30. Letter from Johann Ludwig Schulze to Christoph Samuel John, 19 August 1790.

AFSt/M 1 C 33a : 87. Christoph Samuel John to Gottlieb Friedrich Stoppelberg, 15 September 1791.

Museum für Naturkunde (MfN), Berlin

Bestand Zoologisches Museum (ZMB)

Fischsammlung Bloch

8794. Specimen of *Cyprinus fimbriatus.*

32625. Specimen of *Salmo tumbil.*

Historische Arbeitsstelle

VIII/423. Originalzeichnungen for Marcus Élieser Bloch, *Allgemeine Naturgeschichte der Fische* (Berlin, 1782–1795).

VIII/424. Originalzeichnungen for Marcus Élieser Bloch, *Allgemeine Naturgeschichte der Fische* (Berlin, 1782–1795).

Bestand Gesellschaft Naturforschender Freunde (GNF)

S. Bloch, Tagebuch 1. Entry on visit to Marcus Élieser Bloch's collection.

S. Bloch, Tagebuch 6, 67 B 3102 a-g. Copy of letter from GNF members to King's secretary, undated.

Staatsbibliothek zu Berlin (SBB), Berlin

Sammlung Darmstaedter Weltreisen 1768

Banks, Sir Joseph. Letter from Joseph Banks to Marcus Élieser Bloch, 24 June 1791.

Universitätsbibliothek Leipzig (UL), Leipzig

Autographensammlung

ASL 213. Letter from Marcus Élieser Bloch to Johann Hermann, 2 February 1792.

1.1.3 The Netherlands

Koninklijke Bibliotheek (KB), The Hague

Veilingcatalogi en Boekhandelscatalogi

Verzcat 4612. Annotated copy of catalogue for 1778 auction of Laurens Theodorus Gronovius' library.

Stadsarchief Amsterdam (SA), Amsterdam

Archief van de Notarissen

344 Salomon Dorper, 10695. Inventory of possessions of Peter Artedi.

Universiteitsbibliotheek Amsterdam (UBA), Amsterdam

Handschriften

hs. III C 22/23. Gessner-Platter albums.

Universiteitsbibliotheek Leiden (UBL), *Leiden*

Correspondence of Laurens Theodorus Gronovius

BPL246. Laurens Theodorus Gronovius to Arnout Vosmaer, 28 December 1751.

1.1.4 Sweden

Kungliga Biblioteket (KBS), *Stockholm*

Handskriftssamling bokhandlare L. De Vylder

MS X1002. Letter from Peter Artedi to his relatives, September 1735.

Stockholms Universitetsbibliotek (SUB), *Stockholm*

Bergius Library

H.VII.8.1.n.4. Peter Artedi, *Idea Institutionum Trichozoologiae.*

Uppsala Universitetsbibliotek (UUB), *Uppsala*

Early Printed Books and Special Collections

D 82 a. Peter Artedi. *Kårt Förtekning på de Träen, Buskar åg Örter, såmm wäxa sponté wid Nordmalings Prästebord äller i de närmaste byar där åmmkring Äfter Dänn Alldra-simplaste åg Klaraste Methoden i årdning satte av Petro Arctædio A:o 1729 in Februario.*

Linnaean Correspondence (Accessed via http://www.alvin-portal.org)

L4202. Letter from Carl Linnaeus to Johan Ernst Gunnerus, 4 March 1769.

L0217. Letter from Paul Heinrich Gerhard Möhring to Carl Linnaeus, 4 October 1737.

L0210. Letter from Christian Gottlieb Ludwig to Carl Linnaeus, 22 October 1737.

L2902. Letter from Alexander Garden to Carl Linnaeus, 12 April 1762.

L2656. Letter from Antoine Gouan to Carl Linnaeus, 8 January 1760.

Wallers Handskriftssamling (Accessed via http://www.alvin-portal.org)

Ms. se-00136. Application to stipend by Peter Artedi, 22 May 1734.

1.1.5 United States

Library of Congress (LC), *Washington D.C.*

QL618.15 A78 1735. Peter Artedi, *Manuscriptum ichthyologicum quod Petrus Artedi elaboravit in usum Thesauri Sebani* (accessed via https://www.loc.gov/item/2018656155/).

1.2 *Printed Works before 1828*

d'Argenville, Antoine-Joseph Dézallier. *La conchyliologie, ou, Histoire naturelle des coquilles de mer, d'eau douce, terrestres et fossils*. Paris: Guillaume de Bure fils aîné, 1780.

Artedi, Peter. *Ichthyologia, sive opera omnia de piscibus*. Leiden: Conradum Wishoff, 1738.

Bacon, Francis. *Instauratio magna*. London: John Bill, 1620.

Bartholin, Thomas. *Historiarum anatomicarum rariorum*. Amsterdam: Joannem Henrici, 1654.

Belon, Pierre. *De aquatilibus, libri duo: cum eiconibus ad vivam ipsorum effigiem, quoad ejus fieri potuit, expressis*. Paris: Carolum Stephanum, 1553.

Birch, Thomas. *The History of the Royal Society for Improving of Natural Knowledge*, 4 vols. London: A. Millar, 1756–1757.

Bloch, Marcus Élieser. *Medicinische Bemerkungen, Nebst einer Abhandlung vom Pyrmonter Augenbrunnen*. Berlin: Christian Friedrich Himburg, 1774.

Bloch, Marcus Élieser. *Allgemeine Naturgeschichte der Fische*, 12 vols. Berlin: Hesse, Realschule, J. Morino, 1782–1795.

Bloch, Marcus Élieser. *Ichtyologie, ou, Histoire naturelle, générale et particulière des poissons*, 12 vols. Translated by Jean Charles Thibault de Laveaux. Berlin: François de la Garde, 1785–1797.

Bonnaterre, Pierre Joseph. *Tableau encyclopédique et méthodique des trois règnes de la nature: Ichthyologie*. Paris: Mme Veuve Agasse, 1788.

Boyle, Robert. "New Pneumatical Experiments About Respiration." *Philosophical Transactions of the Royal Society of London* 5, no. 62 (1670): 2011–2031.

Catesby, Mark. *The Natural History of Carolina, Florida and the Bahama Islands*, 2 vols. London: W. Innys and R. Manby, 1729–1747.

Chacón, Alfonso. *Historia iutruisque belli datici a Traiano Caesare gesti quae in columna euisdem Romae visuntur, collecta*. Rome: Franciscum Zanettum & Bartholomaeum Tosium socios, 1576.

Coquebert de Montbret, Antoine-Jean. "Éloge de Monsieur Bloch." *Rapport général des travaux de la Société philomatique de Paris* 4 (1800): 136–156.

Cuvier, Georges. *Tableau historique des progrès de l'ichtyologie depuis son origine jusqu'à nos jours*. Paris & Strasbourg: F.G. Levrault, 1828.

"D.M.E. Bloch, *Oecnomische Naturgeschichte der Fische Teutschlands*." *Allgemeine deutsche Bibliothek* 53, no. 2 (1783): 582–583.

Der Anzeiger: Ein Tagblatt zum Behuf der Justiz, der Polizey und aller bürgerlicher Gewerbe, wie auch zur freyen gegenseitigen Unterhaltung der Leser über gemeinnützige Gegenstände aller Art 19, no. 19–20, 1792.

Figulus, Carolus. *Ichthyologia seu dialogus de piscibus*. Cologne: Eucharius Cervicornus, 1540.

"Francisci Willughbeii Armig. de *Historia piscium* libri quatuor." *Philosophical Transactions of the Royal Society of London*, 15, no. 178 (1685): 1301–1308.

Gessner, Conrad. *Historia animalium liber IIII qui est de piscium & aquatilium animantium natura*. Zurich: Christopher Froschauer, 1558.

Gessner, Conrad. *Fischbuch*, trans. Conrad Forrer. Zurich: Christopher Froschauer, 1563.

Goeree, Willem. *Verligterie-kunde, of regt gebruik der water-verwen*. Amsterdam: Andries van Damme, 1705.

Grew, Nehemiah. *Musaeum Regalis Societatis, or, A catalogue & description of the natural and artificial rarities belonging to the Royal Society and preserved at Gresham Colledge*. London: Hugh Newman, 1681.

Gronovius, Johan Frederik. "A Method of preparing Specimens of Fish, by drying their Skins, as practised by John Frid. Gronovius M.D. in Leyden." *Philosophical Transactions of the Royal Society of London* 42, no. 463 (1742): 57–58.

Gronovius, Laurens Theodorus. *Museum ichthyologicum, sistens piscium indigenorum et quorumdam exoticorum*. Leiden: Theodoor Haak, 1754.

Gronovius, Laurens Theodorus. *Zoophylacium Gronovianum: exhibens animalia quadrupeda, amphibia, pisces, insecta, vermes, mollusca, testacea, et zoophyta*. Leiden: Theodoor Haak, 1781.

Hariot, Thomas. *A Briefe and True Report of the New Found Land of Virginia*. Frankfurt: Johann Wechel, 1590.

Harvey, William. *Exercitationes de generatione animalium*. London: Du Gardianis, 1651.

Intelligenzblatt der Allgemeinen Literatur-Zeitung vom Jahre 1792. Leipzig: Johann Gottfried Müllerischen Büchhandler, 1792.

Intelligenzblatt der Allgemeinen Literatur-Zeitung vom Jahre 1797. Leipzig: Johann Gottfried Müllerischen Büchhandler, 1797.

Jaucourt, Louis de. "Ichthyologie." *Encyclopédie ou Dictionnaire raisonné des sciences, des arts et des métiers*, vol. 8, edited by Denis Diderot and Jean le Rond d'Alembert, 483–485. Neufchâtel: Samuel Faulche, 1765.

John, Christoph Samuel. "Einige Nachrichten von Trankenbar auf der Küste Koromandel. Aus einem Briefe von dem Missionarius Hrn John an Hrn Doktor Bloch in Berlin." *Berlinische Monatschrift* 20 (1792): 585–596.

John, Christoph Samuel. "Fragen des Herrn Professor Forster in Halle, an die Missionarien in Trankenbar, und Herrn Johns Antworten darauf." *Neuere Geschichte der evangelischen Missionsanstalten zu Bekehrung der Heiden in Ost-Indien* 4, no. 43 (1793): 648–667.

John, Christoph Samuel. "Einige Nachrichten von der Küste Koromandel. Auszug eines Schreibens des Hrn Missionarius C.S. John an Hrn D. Bloch in Berlin." *Berlinische Monatschrift* 24 (1794): 585–596.

Klein, Jacob Theodoor. "Peter Artedi, *Ichthyologia*." *Bibliothèque germanique ou Histoire littéraire de l'Allemagne et des pays du Nord*, vol. 45, edited by Paul-Emile de Mauclerc, 143–159. Amsterdam: Pierre Humbert et fils, 1739.

Klein, Jacob Theodoor. *Historiae piscium naturalis*. Gdansk: Thomas Johann Schreiber, 1740–1749.

Kohl, Johann Peter. "Opera posthuma ichtyologica." *Hamburgische Berichte von den neuesten Gelehrten Sachen* 16 (1738): 131–132.

Krünitz, Johann Georg. *Oeconomische Encyclopädie*. Berlin: Joachim Pauli, 1773–1796.

Lacépède, Bernard Germain de. *Histoire naturelle des poissons*, 12 vols. Paris: Plassan, 1798–1803.

Leydse Courant, "Advertisement for books of Conrad Wishoff." May 7, 1738.

Linnaeus, Carl. *Systema naturae*, ed. 1. Leiden: Theodoor Haak, 1735.

Linnaeus, Carl. *Fundamenta botanica*. Amsterdam: Salomon Schouten, 1736.

Linnaeus, Carl. *Critica botanica*. Leiden: Conrad Wishoff, 1737.

Linnaeus, Carl. *Genera plantarum*. Leiden: Conrad Wishoff, 1737.

Linnaeus, Carl. *Systema naturae*, ed. 10. Stockholm: Lars Salvius, 1758.

Magazin des Buch- und Kunsthandels, welches zum Besten der Wissenschaften und Künste von den dahin gehörigen Neuigkeiten Nachtricht giebt. Leipzig: Johann Gottlob Immanuel Breitkopf, no. 7, 1782.

Marschalk, Nicolaus. *Historia aquatilium latine ac grace cum figuris*. Rostock: Nicolaus Marschalk, 1517.

Martini, Friedrich Wilhelm Heinrich. "Gesetze der Hiesigen Gesellschaft," *Beschäftigungen der Berlinischer Gesellschaft Naturforschender Freunde* 1, no. 1 (1775): 27–40.

Mascall, Leonard. *A Booke of Fishing with Hooke & Line, and of All Other Instruments There-unto Belonging*. London: John Wolfe, 1590.

Nicolai, Christoph Friedrich. *Beschreibung der Königlichen Residenzstädte Berlin und Potsdam*, vol. 2. Berlin: Friedrich Nicolai, 1779.

Nieuhof, John. *Gedenkweerdige Brasiliaense Zee- en Lant- Reize*. Amsterdam: J. van Meurs, 1682.

Nozeman, Cornelis and Johann Christiaan Sepp. *Afbeeldingen en beschrijvingen van in- en uitlandsche visschen M.E. Bloch; gevolgd naar het Hoogduitsch*. Zaltbommel, 1804.

"Petri Artedi." *Neue Zeitungen von gelehrten Sachen* 24, no. 1 (1738): 347–351.

"Petri Artedi." *Nova acta eruditorum*, edited by Friedrich Otto Mencke, 652–657. Leipzig: Johann Friedrich Gleditsch, 1741.

Piso, Willem and Georg Marcgraf. *Historia naturalis Brasiliae*. Amsterdam: Ludovicus Elzevier, 1648.

Ray, John. "An account of the dissection of a Porpess, promised numb. 74; made, and communicated in a letter of Sept. 12 1671, by the learned Mr. John Ray, having there in obser'd some things omitted by Rondeletius." *Philosophical Transactions of the Royal Society of London* 6, no. 76 (1671): 2274–2279.

Ray, John. *Observations Topographical, Moral and Physiological, Made in a Journey through Part of the Low-Countries, Germany, Italy and France*. London: John Martyn, 1673.

Ray, John. *A Collection of English words, not generally used ... in two alphabetical catalogues, ... northern ... [and] southern counties, with catalogues of English birds and fish, and an account of preparing ... metals and minerals*. London: Thomas Burrell, 1674.

Ray, John. *The Wisdom of God Manifested in the Works of the Creation*. London: Samuel Smith, 1691.

Ray, John. *Synopsis methodica avium & piscium*. London: William Derham, 1713.

Réaumur, René-Antoine Ferchault de. "Divers Means for Preserving from Corruption Dead Birds, Intended to Be Sent to Remote Countries, So That They May Arrive There in a Good Condition. Some of the Same Means May be Employed for Preserving Quadrupeds, Reptiles, Fishes, and Insects, by M. de Reaumur, F.R.S. and Memb. Royal. Acad. Sc. Paris. Translated from the French by Phil. Hen. Zollman, Esq; F.R.S." *Philosophical Transactions of the Royal Society of London* 45, no. 487 (1748): 304–320.

Rondelet, Guillaume. *L'Histoire entière des poissons*. Lyon: Macé Bonhomme, 1558.

Salter, James. *A Catalogue of the Rarities, to be Seen at Don Saltero's Coffee House*, vol. 10. London: s.n., 1731.

Salviani, Hippolyto. *Aquatilium animalium historiae liber primus*. Rome: H. Salviani, 1554.

Samling af testamenten och författningar om stipendien. Uppsala: Johan Fr. Edman, 1795.

Schoenevelde, Stephan. *Ichthyologia et nomenclaturae animalium marinorum, fluviatilium, lacustrium*. Hamburg: ex Bibliopolio Heringiano, 1624.

von Schreber, Johann Christian. *Die Säughtiere in Abbildungen nach der Natur mit Beschreibungen*. Erlangen: Wolfgang Walther, 1774–1804.

Seba, Albertus. *Locupletissimi rerum naturalium thesauri accurate description et iconibus artificiosissimis expressio*. Amsterdam: apud Wetstenium, & Gul. Smith, & Janssonio-Waesbergios, 1734–1765.

Skippon, Phillip. *An Account of a Journey Made Thro. Part of the Low Countries, Germany, Italy and France*. London: Churchill, 1732.

Sloane, Hans. *A voyage to the islands Madera, Barbados, Nieves, S. Christophers and Jamaica with the Natural History of the Herbs, and Trees, Four-footed Beasts, Fishes, Birds, Insects, Reptiles &c. of the Last of Those Islands*, vol. 1. London: B.M. for the Author, 1707.

Sprat, Thomas. *The History of the Royal-Society of London for the Improving of Natural Knowledge*. London: John Martyn, 1667.

Terry, Edward. *A Voyage to East-India*. London: J. Wilkie, 1655.

Turgot, Étienne-François. *Mémoire instructif sur la manière de rassembler, de preparer, de conserver, et d'envoyer les diverses curiosités d'histoire naturelle*. Lyon: chez Jean Marie Bruyset, 1758.

Walbaum, Johann Julius. *Petri Artedi Renovati*. Greifswald: Anton Ferdinand Röse, 1789.

Wilkins, John. *An Essay Towards a Real Character, and a Philosophical Language*. London: S. Gellibrand and John Martyn, 1668.

Willughby, Francis. *Francisci Willughbeii De Middleton in agro Warwicensi, Armigeri, e Regia Societate, Ornithologiae libri tres*, edited by John Ray. London: John Martyn, 1676.

Willughby, Francis. *The ornithology of Francis Willughby of Middleton in the county of Warwick Esq*, translated and edited by John Ray. London: John Martyn, 1678.

Willughby, Francis and John Ray. *Francisci Willughbeii Armig. De historia piscium libri quatuor*. Oxford: Sheldonian Theatre, 1686.

1.3 *Editions*

Aristotle, *Histories of Animals in Ten Books*. Edited and translated by Richard Cresswell. London: George Bell, 1878.

Bacon, Francis. *The Instauratio Magna Part II: Novum Organum and Associated Texts*. Edited and translated by Graham Rees and Maria Wakely. Oxford: Clarendon Press, 2004.

Cantimpré, Thomas of. *De natura rerum*. Edited by Helmut Boese. Berlin: De Gruyter, 1973.

Cuvier, Georges. *Historical Portrait of the Progress of Ichthyology: From its Origins to our Own Time*. Edited and translated by Theodore W. Pietsch. Baltimore: Johns Hopkins University Press, 1995.

Goethe, Johann Wolfgan von. *Goethes Briefwechsel mit den Gebrüdern Humboldt (1767–1832)*. Edited by F.Th. Bratranek. Leipzig: F.A. Brockhaus, 1876.

Pliny, *Natural History*, 10 vols. Edited and translated by H. Rackham. Cambridge, Mass.: Harvard University Press, 1938–1952.

Ray, John. *The Correspondence of John Ray*. Edited by Edwin Lankester. London: The Ray Society, 1848.

Ray, John. *Further Correspondence of John Ray*. Edited by Robert W. Theodore Gunther. London: The Ray Society, 1928.

Willughby, Cassandra. "An Account of an Elizabethan Family: The Willoughbys of Wollaton by Cassandra Willoughby, 1670–1735." Edited by Jo Ann Hoeppner Moran Cruz, *Royal Historical Society Camden Fifth Series* 55 (2018): 67–258.

2 Secondary Sources

Abeele, Baudouin van den. "Conrad Gessner als Leser mittelalterlicher Enzyklopädien." In Leu and Opitz, *Conrad Gessner*, 15–28.

Achbari, Azadeh. "Rulers of the Winds: How Academics Came to Dominate the Science of Weather." PhD diss., Vrije Universiteit Amsterdam, 2017.

Ahnert, Ruth. "Maps versus Networks." In Raymond and Moxham, *News Networks in Early Modern Europe*, 130–157.

Aili, Hans and Theodore W. Pietsch. "Jacob Theodor Klein's Critique of Peter Artedi's *Ichthyologia* (1738)." *Svenska Linnésällskapets Årsskrift* 97 (2014): 39–84.

Aili, Hans and Theodore W. Pietsch. *Peter Artedi: Reformer of 18th Century Zoology*, vol. 1. Stockholm: Stockholm University Press, 2024.

Anstey, Peter R. "Philosophy of Experiment in Early Modern England: The Case of Bacon, Boyle and Hooke." *Early Science and Medicine* 19, no. 2 (2014): 103–132.

Aranowsky, Leah. "On Drawing Dead Fish." *Environmental History* 21, no. 3 (2016): 543–551.

Argeloo, Marc. *Natuuramnesie: Hoe we vergeten zijn hoe de natuur er vroeger uitzag*. Amsterdam: Atlas Contact, 2022.

Ashworth, William B. "Natural History and the Emblematic Worldview." In *Reappraisals of the Scientific Revolution*, edited by David C. Lindberg and Robert S. Westman, 303–332. Cambridge: Cambridge University Press, 1990.

Ashworth, William B. "Emblematic Natural History of the Renaissance." In Jardine, Secord and Spary, *Cultures of Natural History*, 17–37.

Azzolini, Monica. "Talking of Animals: Whales, Ambergris, and the Circulation of Knowledge in Seventeenth-Century Rome." *Renaissance Studies* 31, no. 2 (2017): 297–318.

Balcombe, Jonathan. *What A Fish Knows*. London: Oneworld, 2019.

Balfe, Thomas, Joanna Woodall and Claus Zittel. "Introduction: The Lives of *Ad vivum*." In *Ad vivum? Visual Materials and the Vocabulary of Life-Likeness in Europe before 1800*, edited by Thomas Balfe, Joanna Woodall and Claus Zittel, 1–43. Leiden: Brill, 2019.

Barnett, Lydia. "Showing and Hiding: The Flickering Visibility of Earth Workers in the Archives of Earth Science." *History of Science* 58, no. 3 (2020): 245–274.

Becker, Marcus, Eva Dolezel, Meike Knittel, Diana Stört and Sarah Wagner, eds. *The Berlin Kunstkammer: Collection History in Object Biographies from the 16th to the 21st Century*. Petersberg: Michael Imhof Verlag, 2023.

Beltrame, Tiziana N., Elena Canadelli and Luca Tonetti. "The Natures of Digital Practices: People, Objects, and Data Mobilities in Natural History Collections." *Nuncius* 39, no. 3 (2024): 741–758.

Bennema, Floris P. and Adriaan D. Rijnsdorp. "Fish Abundance, Fisheries, Fish Trade and Consumption in Sixteenth-Century Netherlands as Described by Adriaen Coenen." *Fisheries Research* 161 (2015): 384–399.

Bennett, Jim and Rebekah Higgitt, eds. *London 1600–1800: Communities of Natural Knowledge and Artificial Practice*, special issue of *British Journal for the History of Science* 52, no. 2 (2019): 183–343.

Bergvelt, Ellinoor and Renée Kistemaker, eds. *De Wereld Binnen Handbereik: Nederlandse Kunst- en Rariteitenverzamelingen, 1585–1735*. Zwolle: Waanders, 1992.

Bergvelt, Ellinoor and Debora J. Meijers, eds. *Teyler's Foundation in Haarlem and Its 'Book and Art Room' of 1779*. Leiden: Brill, 2020.

Berkeley, Edmund Jr. and Dorothy Smith Berkeley, eds. *The Correspondence of John Bartram 1734–1777*. Gainesville: University Press of Florida, 1992.

Bertolini Meli, Domenico. *Visualizing Disease: The Art and History of Pathological Illustrations*. Chicago: Chicago University Press, 2017.

Blair, Ann. "Annotating and Indexing Natural History." In Frasca-Spada and Jardine, *Books and the Sciences in History*, 69–89.

Blair, Ann and Kaspar von Greyerz, eds. *Physico-Theology: Religion and Science in Europe, 1650–1750*. Baltimore: Johns Hopkins University Press, 2020.

Blanken, Kerrewin van. "Earthquake Observations in the Age Before Lisbon: Eyewitness Observation and Earthquake Philosophy in the Royal Society, 1665–1755." *Notes and Records of the Royal Society of London* 76, no. 1 (2020): 27–48.

Bleichmar, Daniela. "Seeing the World in a Room: Looking at Exotica in Early Modern Collections." In *Collecting Across Cultures: Material Exchanges in the Early Modern Atlantic World*, edited by Daniela Bleichmar and Peter C. Mancall, 15–30. Philadelphia: University of Pennsylvania Press, 2011.

Bleichmar, Daniela. *Visible Empire: Botanical Expeditions & Visual Culture in the Hispanic Enlightenment*. Chicago: Chicago University Press, 2012.

Bleichmar, Daniela and Peter C. Mancall. "Introduction." In Bleichmar and Mancall, *Collecting Across Cultures*, 1–14.

Blunt, Wilfrid. *The Compleat Naturalist: A Life of Linnaeus*. London: Collins, 1971.

Boeseman, Marinus. "The Vicissitudes and Dispersal of Albertus Seba's Zoological Specimens." *Zoologische Mededelingen* 44, no. 13 (1970): 177–210.

Böhme, Katrin. "Die Gesellschaft Naturforschender Freunde zu Berlin: Bestand und Wandel einer gelehrten Gesellschaft Ein Überblick." *Berichte zur Wissenschaftsgeschichte* 24, no. 4 (2001): 271–283.

Böhme, Katrin. *Gemeinschaftsunternehmen Naturforschung: Modifikation und Tradition in der Gesellschaft Naturforschender Freunde zu Berlin 1773–1906*. Stuttgart: Franz Steiner, 2005.

Bouwman, André, Ed van der Vlist, Berry Dongelmans and Paul Hoftijzer, eds. *Stad van Boeken: Handschrift en Druk in Leiden, 1260–2000*. Leiden: Primavera, 2008.

Bruhn, Matthias. "Beyond the Icons of Knowledge: Artistic Styles and the Art History of Scientific Imagery." In *The Technical Image: A History of Styles in Scientific Imagery*, edited by Horst Bredekamp, Vera Dünkel and Birgit Schneider, 36–45. Chicago: Chicago University Press, 2015.

Buis, Alena, Christi Spain-Savage and Myra E. Wright. "Attending to Fishwives: Views from Seventeenth-Century London and Amsterdam." In *Mapping Gendered Routes and Spaces in the Early Modern World*, edited by Merry E. Wiesner-Hanks, 177–200. Farnham: Ashgate, 2015.

Burnett, Graham D. *Trying Leviathan: The Nineteenth-Century New York Court Case That Put the Whale on Trial and Challenged the Order of Nature*. New Jersey: Princeton University Press, 2007.

Birkhead, Tim, ed. *Virtuoso by Nature: The Scientific Worlds of Francis Willughby* FRS *(1635–1672)*. Leiden: Brill, 2016.

Birkhead, Tim, ed. *The Wonderful Mr. Willughby: The First True Ornithologist*. London: Bloomsbury, 2018.

Birkhead, Tim, Paul J. Smith, Meghan Doherty and Isabelle Charmantier. "Willughby's Ornithology." In Birkhead, *Virtuoso by Nature*, 268–304.

Brigham, David R. "Mark Catesby and The Patronage of Natural History in the First Half of the Eighteenth Century." In *Empire's Nature: Mark Catesby's New World Vision*, edited by Amy Meyers and Margaret Beck Pritchard, 91–146. Chapel Hill: University of North Carolina Press, 1998.

Broberg, Gunnar. "Petrus Artedi in his Swedish Context." In *Proceedings of the Fifth Congress on European Ichthyology*, edited by Sven O. Kullander and Bo Fernholm, 11–15. Stockholm: Swedish Museum of Natural History, 1987.

Broberg, Gunnar. *Mannen som ordnade naturen: En biografi över Carl von Linné*. Stockholm: Natur & Kultur, 2019.

Cambefort, Yves. "How General are Genera? The Genus in Systematic Zoology." In *The Oxford Handbook of Generality in Mathematics and the Sciences*, edited by Karine Chemla, Renaud Chorlay and David Rabouin, 257–285. Oxford: Oxford University Press, 2016.

Carhart, Michael C. "Historia Literaria and Cultural History: From Mylaeus to Eichhorn," in *Momigliano and Antiquarianism: Foundations of the Cultural Sciences*, edited by Peter Miller, 184–206. Toronto: University of Toronto Press, 2007.

Ceríaco, Luis and Mariana Marques. "Peixes Em 'Herbário.' Uma Técnica Científico-Museológica do Século XVIII." Conference Paper, Congresso Luso-Brasileiro de História das Ciências (October 2011): 1204–1219.

Chakrabarti, Pratik. *Materials and Medicine Trade, Conquest and Therapeutics in The Eighteenth Century*. Manchester: Manchester University Press, 2023.

Charmantier, Isabelle. "Emblematics in Ornithology in the Sixteenth and Seventeenth Centuries." *Emblematica* 18 (2010): 79–109.

Charmantier, Isabelle. "Carl Linnaeus and the Visual Representation of Nature." *Historical Studies in the Natural Sciences* 41, no. 4 (2011): 365–404.

Charmantier, Isabelle and Staffan Müller-Wille. "Lists as Research Technologies." *Isis* 103, no. 4 (2012): 743–752.

Charmantier, Isabelle and Staffan Müller-Wille. "Natural History and Information Overload: The Case of Linnaeus." *Studies in History and Philosophy of Science*

Part C: Studies in History and Philosophy of Biological and Biomedical Sciences 43, no. 1 (2012): 4–15.

Charmantier, Isabelle and Staffan Müller-Wille. "Carl Linnaeus's Botanical Paper Slips (1767–1773)." *Intellectual History Review* 24, no. 2 (2014): 215–238.

Charmantier, Isabelle, Dorothy Johnston and Paul J. Smith. "The Legacies of Francis Willughby." In Birkhead, *Virtuoso by Nature*, 360–385.

Chen, Jessie Wei-Hsuan. "A Woodblock's Career: Transferring Visual Botanical Knowledge in the Early Modern Low Countries." *Nuncius* 35, no. 1 (2020): 20–63.

Christensson, Jakob. *Ichtyologia. Det vill säga alla verk om fiskarna*. Stockholm: Kungliga Skogs- och Lantbruksakademien, 2022.

Cooper, Alix. "The Indigenous versus the Exotic: Debating Natural Origins in Early Modern Europe." *Landscape Research* 28, no. 1 (2003): 51–60.

Cooper, Alix. *Inventing the Indigenous: Local Knowledge and Natural History in Early Modern Europe*. Cambridge: Cambridge University Press, 2007.

Cooper, Alix. "Homes and Households." In Park and Daston, *Cambridge History of Science* vol. 3, 224–237.

Cram, David. "Francis Willughby and John Ray on Words and Things." In Birkhead, *Virtuoso by Nature*, 244–267.

Curry, Helen Anne, Nicholas Jardine, James Andrew Secord and Emma C. Spary, eds. *Worlds of Natural History*. Cambridge: Cambridge University, 2019.

Dackerman, Susan. "Introduction: Prints as Instruments." In *Prints and the Pursuit of Knowledge*, edited by Susan Dackerman, 19–35. Cambridge, Mass.: Harvard Art Museums, 2011.

Daston, Lorraine. "Description by Omission: Nature Enlightened and Obscured." In *Regimes of Description: In the Archive of the Eighteenth Century*, edited by John Bender and Michael Marrinan, 11–24. Stanford: Stanford University Press, 2005.

Daston, Lorraine. "Science Studies and the History of Science." *Critical Inquiry* 35, no. 4 (2009): 798–813.

Daston, Lorraine. "Super-Vision: Weather Watching and Table Reading in the Early Modern Royal Society and Académie Royale des Sciences." *Huntington Library Quarterly* 78, no. 2 (2015): 187–215.

Daston, Lorraine. "Epistemic Images." In *Vision and Its Instruments: Art, Science, and Technology in Early Modern Europe*, edited by Alina Payne, 13–35. Pennsylvania State University Press: Philadelphia, 2015.

Daston, Lorraine and Katherine Park. *Wonders and the Order of Nature, 1150–1750*. New York: Zone Books, 1998.

Daston, Lorraine and Peter Galison. *Objectivity*. New York: Zone Books, 2007.

Davis, Peter "Collecting and Preserving Fish: a Historical Perspective." In MacGregor, *Naturalists in the Field*, 149–165.

Dear, Peter. "Totius in Verba: Rhetoric and Authority in the Early Royal Society." *Isis* 76, no. 2 (1985): 145–161.

Dear, Peter. *Discipline and Experience: The Mathematical Way in the Scientific Revolution.* Chicago: Chicago University Press, 1995.

Dear, Peter. "The Meanings of Experience." In Park and Daston, *Cambridge History of Science*, vol. 3, 106–131.

Dear, Peter. "Science is Dead; Long Live Science." *Osiris* 27, no. 1 (2012): 37–55.

Delbourgo, James. "Divers Things: Collecting the World Underwater." *History of Science* 49, no. 2 (2011): 149–186.

Delbourgo, James. *Collecting the World: The Life and Curiosity of Hans Sloane.* London: Allen Lane, 2017.

Delbourgo, James and Staffan Müller-Wille. "Introduction: Listmania," *Isis* 103, no. 4 (2012): 710–715.

"Description of Isaac van Duynen's 'Stilleven met vissen op een tafel'." *Hoogsteder Journal* 3 (1997): 21.

Dietz, Bettina. "Linnaeus' Restless System: Translation as Textual Engineering in Eighteenth-Century Botany." *Annals of Science* 73, no. 2 (2014): 143–156.

Dolezel, Eva. "Lehrreiche Unterhalting oder Wissenschaftliche Hülfsmittel? Die Berliner Kunstkammer um 1800. Eine Sammlung am Schnittpunkt Zweier Musealer Konzepte." *Jahrbuch der Berliner Museen* 46 (2004): 147–160.

Dolezel, Eva. *Der Traum vom Museum: die Kunstkammer im Berliner Schloss um 1800: eine museumsgeschichtliche Verortung.* Berlin: Gebr. Mann, 2019.

Dolezel, Eva, Rainer Godel, Andreas Peča and Holger Zaunstöck, eds. *Ordnen – Vernetzen – Vermitteln. Kunst- und Naturalienkammern der Frühen Neuzeit als Lehr- und Lernort.* Stuttgart: Wissenschaftliche Verlagsgeselschaft, 2018.

Dolezel, Eva, Meike Knittel and Diana Stört. "Around 1800: The *Kunstkammer* in Transition." In *The Berlin Kunstkammer*, edited by Marcus Becker, Eva Dolezel, Meike Knittel, Diana Stört and Sarah Wagner, 174–177. Petersberg: Michael Imhof Verlag, 2023.

Driessen-van het Reve, Jozien J. *De Kunstkamera van Peter de Grote: De Hollandse Inbreng, Gereconstrueerd uit Brieven van Albert Seba en Johann Daniel Schumacher, uit de Jaren 1711–1752.* Hilversum: Verloren, 2006.

Dupré, Sven and Christoph Lüthy. "Introduction: Silent Messengers. The World of Goods and the Circulation of Knowledge in the Early Modern Netherlands." In *Silent Messengers: The Circulation of Material Objects of Knowledge in the Early Modern Low Countries*, edited by Sven Dupré and Christoph Lüthy, 1–12. Berlin: LIT, 2011.

Eddy, Matthew D. "Tools for Reordering: Commonplacing and the Space of Words in Linnaeus' *Philosophia Botanica*." *Intellectual History Review* 20, no. 2 (2010): 227–252.

Egmond, Florike. "A Collection within a Collection: Rediscovered Animal Drawings from the Collections of Conrad Gessner and Felix Platter." *Journal for the History of Collections* 25, no. 2 (2013): 149–170.

Egmond, Florike. *Eye for Detail: Images of Plants and Animals in Art and Science 1500–1630*. London: Reaktion Books, 2017.

Egmond, Florike. "On Northern Shores: Sixteenth-Century Observations of Fish and Seabirds (North Sea and North Atlantic)." In MacGregor, *Naturalists in the Field*, 129–148.

Egmond, Florike. "European Exchanges and Communities." In Curry, Jardine, Secord and Spary, *Worlds of Natural History*, 78–93.

Enenkel, Karl A.E. "Die antike Vorgeschichte der Verankerung der Naturgeschichte in Politik und Religion: Plinius' Zoologie und der römische Imperialismus." In Enenkel and Smith, *Zoology in Early Modern Culture*, 13–54.

Enenkel, Karl A.E. and Paul J. Smith, eds. *Zoology in Early Modern Culture: Intersections of Science, Theology, Philology, and Political and Religious Education*. Leiden: Brill, 2014.

Engel, Hendrik. "Some Artedi Documents in the Amsterdam Archives." *Svenska Linnésällskapets Årsskrift* 34 (1951): 51–66.

Farber, Paul Lawrence. *The Emergence of Ornithology as a Scientific Discipline, 1760–1850*. Dordrecht: Springer, 1982.

Feiner, Shmuel. *The Jewish Enlightenment*. Philadelphia: University of Pennsylvania Press, 2002.

Feiner, Shmuel. *The Jewish Eighteenth Century: A European Biography, 1700–1750*. Bloomington: Indiana University Press, 2020.

Findlen, Paula. *Possessing Nature: Museums, Collecting, and Scientific Culture in Early Modern Italy*. Berkeley: University of California Press, 1994.

Findlen, Paula. "The Formation of a Scientific Community: Natural History in Sixteenth-Century Italy." In *Natural Particulars: Nature and the Disciplines in Renaissance Europe*, edited by Anthony Grafton and Nancy Siraisi, 369–400. Cambridge, Mass.: MIT Press, 1999.

Findlen, Paula. "Natural History." In Park and Daston, *Cambridge History of Science*, vol. 3, 436–468.

Findlen, Paula. "Anatomy Theaters, Botanical Gardens, and Natural History Collections." In *ibid.*, 272–289.

Findlen, Paula and Anna Toledano. "The Materials of Natural History." In Curry, Jardine, Secord, and Spary, *Worlds of Natural History*, 151–169.

Fischel, Angela. *Natur im Bild: Zeichnung und Naturerkenntnis bei Conrad Gessner und Ulisse Aldrovandi*. Berlin: Gebr. Mann, 2009.

Fischer, Dorothée. "The Afterlives of Fish Far from Home: (Mis)Representations in the Iconography of Preserved and Printed Pufferfish in 18th-Century Germany." In *Ichthyology in Context (1500–1800)* edited by Paul J. Smith and Florike Egmond, 554–583. Leiden: Brill, 2023.

Fluck, Hans-R. and Albert Scharbach. "Leonhard Baldner – Zu seinem Testament and Nachlassverzeichniss." *Revue d'Alsace* 142 (2016): 283–297.

Frängsmyr, Tore. "Linnaeus in his Swedish Context." In *Contemporary Perspectives on Linnaeus*, edited by John Weinstock, 183–186. Lanham: University Press of America, 1985.

Frängsmyr, Tore. "The Mathematical Philosophy." In Frängsmyr, Heilbron and Rider, *The Quantifying Spirit*, 27–44.

Fransen, Sietske. "Antoni van Leeuwenhoek, His Images and Draughtsmen." *Perspectives on Science* 27, no. 3 (2019): 485–544.

Fransen, Sietske. "Introduction: Translators and Translations of Early Modern Science." In *Translating Early Modern Science*, edited by Sietske Fransen, Niall Hodson and Karl A.E. Enenkel, 1–14. Leiden: Brill, 2017.

Fransen, Sietske, Katherine M. Reinhart and Sachiko Kusukawa. "Copying Images in the Archives of the Early Royal Society." *Word & Image* 35, no. 3 (2019): 256–276.

Frasca-Spada, Marina and Nicholas Jardine, eds. *Books and the Sciences in History*. Cambridge: Cambridge University Press, 2000.

Fyfe, Aileen and Noah Moxham. "Making Public Ahead of Print: Meetings and Publications at the Royal Society, 1752–1892." *Notes and Records of the Royal Society of London* 70, no. 4. (2016): 361–379.

Gaukroger, Stephen. *Descartes' System of Natural Philosophy*. Cambridge: Cambridge University Press, 2002.

Gentilcore, David. *Food and Health in Early Modern Europe: Diet, Medicine and Society, 1450–1800*. London: Bloomsbury, 2016.

Geus, Armin. "Leonhard Baldner: A Strasbourg Fisherman." *Isis* 55, no. 2 (1964): 195–199.

Giere, Peter, Peter Bartsch, and Christiane Quaisser. "From Humboldt to HVac – The Zoological Collections of the Museum für Naturkunde Leibniz Institute for Evolution and Biodiversity Science in Berlin." In *Zoological Collections of Germany: The Animal Kingdom in its Amazing Plenty in Museums and Universities*, edited by Lothar A. Beck, 89–122. Dordrecht: Springer, 2018.

Gieryn, Thomas F. "Boundary-Work and the Demarcation of Science from Non-Science: Strains and Interests in Professional Ideologies of Scientists." *American Sociological Review* 48, no. 6 (1983): 781–795.

Giglioni, Guido. "Learning to Read Nature: Francis Bacon's Notion of Experiential Literacy (*Experientia Literata*)." *Early Science and Medicine* 18, no. 4/5 (2013): 405–434.

Giglioni, Guido, Dana Jalobeanu and Sorana Corneanu, eds. *Francis Bacon and the Reconfiguration of Early Modern Natural History*, special issue of *Early Science and Medicine* 17, no. 1/2 (2012): 1–271.

Goldgar, Anne. *Impolite Learning: Conduct and Community in the Republic of Letters, 1680–1750*. New Haven: Yale University Press, 1995.

Grafton, Anthony. *New Worlds, Ancient Texts: The Power of Tradition and the Shock of Discovery*. Cambridge, Mass.: Harvard University Press, 1995.

Grafton, Anthony. "Philological and Artisanal Knowledge Making in Renaissance Natural History: A Study in Cultures of Knowledge." *History of the Humanities* 3, no. 1 (2018): 39–55.

Graham, Elspeth. "Ways of Being, Ways of Knowing: Fish, Fishing, and Forms of Identity in Seventeenth-Century English Culture." In *Animals and Early Modern Identity*, edited by Pia F. Cuneo, 351–373. Farnham: Ashgate, 2014.

Guerrini, Anita. *Experimenting with Humans and Animals: From Galen to Animal Rights*. Baltimore: Johns Hopkins University Press, 2003.

Guerrini, Anita. "The Material Turn in the History of Life Science: Life Sciences" *Literature Compass* 13, no. 7 (2016): 469–480.

Guerrini, Anita. *The Courtiers' Anatomists: Animals and Humans in Louis XIV's Paris*. Chicago: Chicago University Press, 2016.

Graheli, Shanti, ed. *Buying and Selling: The Business of Books in Early Modern Europe*. Leiden: Brill, 2016.

Greengrass, Mark, and Daisy Hildyard, Christopher D. Preston and Paul J. Smith, "Science on the Move: Francis Willughby's Expeditions." In Birkhead, *Virtuoso by Nature*, 142–226.

Harkness, Deborah E. *The Jewel House: Elizabethan London and the Scientific Revolution*. New Haven: Yale University Press, 2007.

Haskell, Francis and Henrietta McBurney. "The Paper Museum of Cassiano dal Pozzo." *Visual Resources* 14, no. 1 (1998): 1–17.

Heesen, Anke te. "Vom naturgeschichtlichen Investor zum Staatsdiener: Sammler und Sammlungen der Gesellschaft Naturforschender Freunde zu Berlin um 1800." In *Sammeln als Wissen. Das Sammeln und seine wissenschaftsgeschichtliche Bedeutung*, edited by Anke te Heesen and Emma Spary, 62–84. Göttingen: Wallstein, 2001.

Heilbron, John. "Introduction." In *The Quantifying Spirit in the Eighteenth Century*, edited by Tore Frängsmyr, J.L. Heilbron, and Robin E. Rider, 1–23. Berkeley: University of California Press, 1990.

Hellawell, Philippa. "Diving Engines, Submarine Knowledge and the 'Wealth Fetch'd out of the Sea.'" *Renaissance Studies* 34, no. 1 (2019): 78–94.

Hellawell, Philippa. "'The Best and Most Practical Philosophers': Seamen and the Authority of Experience in Early Modern Science," *History of Science* 58, no. 1 (2019): 28–50.

Heller, John L. "Linnaeus's *Bibliotheca Botanica*." *Taxon* 19, no. 3 (1970): 363–411.

Heller, John L. "Linnaeus on Sumptuous Books." *Taxon* 25, no. 1 (1976): 33–52.

Henderson, Felicity. "Translation in the Circle of Robert Hooke." In Fransen, Hodson and Enenkel, *Translating Early Modern Science*, 17–40.

Henderson, Felicity. "Robert Hooke and the Visual World of the Royal Society." *Perspectives on Science* 27, no. 3 (2019): 395–434.

Hendriksen, Marieke M.A. "Animal Bodies between Wonder and Natural History: Taxidermy in the Cabinet and Menagerie of Stadholder Willem V (1748–1806)." *Journal of Social History* 52, no. 4 (2019): 1110–1131.

Hendrikx, Sophia. "Identification of Herring Species in Conrad Gessner's Ichthyological Works: A Case Study on Taxonomy, Nomenclature, and Animal Depiction in the Sixteenth Century." In Enenkel and Smith, *Zoology in Early Modern Culture*, 149–171.

Hendrikx, Sophia. "Gessner's Taxonomical Skill Exhibited in his Discussion of Felchen." In Leu and Opitz, *Conrad Gessner*, 607–638.

Hendrikx, Sophia. "Conrad Gessner's Fish Books (1556–1560): Processing Information in a Rapidly Expanding Field of Knowledge." PhD diss., Leiden University, 2024.

Hochstrasser, Julie Berger. "From the Waters: Fish Still Life." In *The Magic of Things: Still-life Painting, 1500–1800*, edited by Jochen Sander, 185–211. Berlin: Hatje Cantz, 2008.

Hodacs, Hanna, Kenneth Nyberg and Stéphane Van Damme, "Introduction: de-centring and re-centring Linnaeus." In *Linnaeus, Natural History and the Circulation of Knowledge*, edited by Hanna Hodacs, Kenneth Nyberg and Stéphane Van Damme, 1–24. Oxford: Voltaire Foundation, 2018.

Holthuis, L.B. "Albertus Seba's 'Locupletissimi rerum naturalium thesauri …' (1734–1765) and the 'Planches de Seba' (1827–1831)." *Zoologische Mededelingen* 43, no. 19 (1969): 239–252.

Hommel, Karsten. "Physico-Theology as Mission Strategy: Missionary Christoph Samuel John's (1746–1813) Understanding of Nature." In *Halle and the Beginning of Protestant Christianity in India* vol. 3, edited by Andreas Gross, Y. Vincent Kumaradoss, and Heike Liebau, 1115–1134. Halle: Franckesche Stiftungen, 2006.

Hünniger, Dominik. "'Extolled by Foreigners': William Hunter's Collection and the Development of Science and Medicine in Eighteenth-Century Europe." In *William Hunter and the Anatomy of the Modern Museum*, edited by Mungo Campbell, Nathan Flis and María Dolores Sánchez-Jáuregui. New Haven: Yale University Press, 2018.

Hünniger, Dominik. "Visible Labour? Productive Forces and Imaginaries of Participation in European Insect Studies, ca. 1680–1810," *Berichte zur Wissenschaftsgeschichte* 44, no. 2 (2021): 180–210.

Hunter, Michael. *Establishing the New Science: The Experience of the Early Royal Society.* Suffolk: The Boydell Press, 1989.

Hunter, Michael. "A 'College' for the Royal Society: The Abortive Plan of 1667–1668." *Notes and Records of the Royal Society in London* 38, no. 2 (1984): 159–186.

Hunter, Michael. "Robert Boyle and the Early Royal Society: A Reciprocal Exchange in the Making of Baconian Science." *British Journal for the History of Science* 40, no. 1 (2007): 1–23.

Hunter, William. *Wicked Intelligence: Visual Art and the Science of Experiment in Restoration London*. Chicago: University of Chicago Press, 2013.

Hsia, Florence. *Sojourners in a Strange Land: Jesuits and Their Scientific Missions in Late Imperial China*. Chicago: University of Chicago Press, 2009.

Iliffe, Rob. "Material Doubts: Hooke, Artisan Culture and the Exchange of Information in 1670s London." *British Journal for the History of Science* 28, no. 3 (1995): 285–318.

Iliffe, Rob, ed. *Technicians*, special issue of *Notes and Records of the Royal Society of London* 62, no. 1 (2008): 3–148.

Impey, Oliver and Arthur MacGregor, eds. *The Origins of Museums: The Cabinet of Curiosities in Sixteenth- and Seventeenth-Century Europe*. Oxford: Clarendon Press, 1985.

Jackson, J.B.C. "Reefs since Columbus." *Coral Reefs* 16 (1997): 23–32.

Jalobeanu, Dana. "Disciplining Experience: Francis Bacon's Experimental Series and the Art of Experimenting." *Perspectives on Science* 24, no. 3 (2016): 324–342.

Jardine, Lisa. "*Experientia literata* or *Novum organum*? The Dilemma of Bacon's Scientific Method." In *Francis Bacon's Legacy of Texts: 'The Art of Discovery Grows with Discovery'*, edited by William A. Sessions, 47–67. New York: Ams Press, 1990.

Jardine, Nicholas, James Secord and Emma C. Spary, eds. *Cultures of Natural History*. Cambridge: Cambridge University Press, 1996.

Jeyaray, Daniel. "Mission Reports from South India and Their Impact on the Western Mind: The Tranquebar Mission of the Eighteenth Century." In *Converting Colonialism: Visions and Realities in Mission History, 1706–1914*, edited by Dana L. Robert, 21–42. Cambridge: William B. Eerdmans, 2008.

Johns, Adrian. *The Nature of the Book: Print and Knowledge in the Making*. Chicago: University of Chicago Press, 2000.

Johns, Adrian. "Miscellaneous Methods: Authors, Societies and Journals in Early Modern England." *British Journal for the History of Science*, 33, no. 2 (2000): 159–186.

Johns, Adrian. "Coffeehouses and Print Shops." In Park and Daston, *Cambridge History of Science*, vol. 3, 320–340.

Johnston, Dorothy. "The Life and Domestic Context of Francis Willughby." In Birkhead, *Virtuoso by Nature*, 1–43.

Jorink, Eric. *Reading the Book of Nature in the Dutch Golden Age, 1575–1715*. Leiden: Brill, 2010.

Jorink, Eric. "Beyond the Lines of Apelles: Johannes Swammerdam, Dutch Scientific Culture and the Representation of Insect Anatomy." *Nederlands Kunsthistorisch Jaarboek (NKJ)/Netherlands Yearbook for History of Art* 61 (2011): 149–183.

Jorink, Eric. "Snakes, Fungi and Insects. Otto Marseus van Schrieck, Johannes Swammerdam and the Theory of Spontaneous Generation." In Enenkel and Smith, *Zoology in Early Modern Culture*, 199–207.

Jorink, Eric. "Insects, Philosophy and the Microscope." In Curry, Jardine, Secord and Spary, *Worlds of Natural History*, 131–147.

Jürgens, Hanco. "Van God's Akker tot Spiegel der Natuur: Veranderende Percepties van de Indiase Natuur in Berichten van Duitse Zendelingen." *De Achttiende Eeuw* 36, no. 2 (2004): 67–91.

Karrer, Christine. "Marcus Elieser Bloch (1723–1799), Sein Leben und die Geschichte seiner Fischsammlung." *Sitzungsberichte der Gesellschaft Naturforschender Freunde zu Berlin* 18 (1978): 129–149.

Kelley, Donald R. "The Problem of Knowledge and the Concept of Discipline." In *History and the Disciplines: The Reclassification of Knowledge in Early Modern Europe*, edited by Donald R. Kelley, 1–12. New York: University of Rochester Press, 1997.

Kemp, Martin. "Style and Non-Style in Anatomical Illustration: From Renaissance Humanism to Henry Gray." *Journal of Anatomy* 216, no. 2 (2010): 192–208.

Kerbert, C. *Peter Artedi 1705–1785, Rede uitgesproken bij de onthulling van zijn gedenkteeken, in den tuin van het Koninklijk Zoölogisch Genootschap "Natura Artis Magistra".* Amsterdam: J.H. de Bussy, 1905.

Kilburn-Toppin, Jasmine. "'A Place of Great Trust to be Supplied by Men of Skill and Integrity': Assayers and Knowledge Cultures in Late Sixteenth- and Seventeenth-Century London." *The British Journal for the History of Science* 52, no. 2 (2019): 197–223.

Koerner [Rausing], Lisbet. *Linnaeus: Nature and Nation*. Cambridge, Mass.: Harvard University Press, 1999.

Koerner [Rausing], Lisbet. "Linnaeus in His Time and Place." In Jardine, Secord and Spary, *Cultures of Natural History*, 145–162.

Krämer, Fabian. *Ein Zentaur in London: Lektüre und Beobachtung in der frühneuzeitlichen Naturforschung*. Korb: Didymos, 2014.

Kreklau, Claudia. "Travel, Technology, and Theory: The Aesthetics of Ichthyology during the Second Scientific Revolution." *German Historical Review* 41, no. 3 (2018): 589–610.

Kusukawa, Sachiko. "Illustrating Nature." In Frasca-Spada and Jardine, *Books and the Sciences in History*, 90–113.

Kusukawa, Sachiko. "The *Historia Piscium* (1686)." *Notes and Records of the Royal Society of London* 54, no. 2 (2000): 179–197.

Kusukawa, Sachiko. "Picturing Knowledge in the Early Royal Society: The Examples of Richard Waller and Henry Hunt." *Notes and Records of the Royal Society* 65, no. 3 (2011): 273–294.

Kusukawa, Sachiko. *Picturing the Book of Nature: Image, Text and Argument in Sixteenth-Century Human Anatomy and Medical Botany*. Chicago: University of Chicago Press, 2012.

Kusukawa, Sachiko. "Drawings of Fossils by Robert Hooke and Richard Waller." *Notes and Records of the Royal Society* 67, no. 2 (2013): 123–138.

Kusukawa, Sachiko. "*Historia Piscium* (1686) and Its Sources." In Birkhead, *Virtuoso by Nature*, 305–334.

Kusukawa, Sachiko. "The Early Royal Society and Visual Culture." *Perspectives on Science* 27, no. 3 (2019): 350–394.

Kusukawa, Sachiko. "*Ad vivum* Images and Knowledge of Nature in Early Modern Europe." In Balfe, Woodall and Zittel, *Ad vivum?*, 89–121.

Kusukawa, Sachiko. "Gessner's History of Nature." In Curry, Jardine, Secord, and Spary, *Worlds of Natural History*, 33–43.

Landsberg, Hannelore. "Eine Fischsammlung aus Tranquebar, die Berliner Gesellschaft Naturforschender Freunde und deren Mitglied Marcus Elieser Bloch." In *Mission und Forschung. Translokale Wissensproduktion zwischen Indien und Europa im 18. und 19. Jahrhundert*, edited by Heike Liebau, 167–179. Halle: Frankesche Stiftungen, 2010.

Langman, Andrew Peter. "'Beyond, both the Old World, and the New': Authority and Knowledge in the Works of Francis Bacon, with Special Reference to the *New Atlantis*." PhD diss., Queen Mary University London, 2007.

Larson, James L. "Linnaeus and the Natural Method." *Isis* 58, no. 3 (1967): 304–320.

Latour, Bruno. *Science in Action: How to Follow Scientists and Engineers Through Society*. Cambridge Mass.: Hardvard University, 1987.

Laube, Stefan. "Privilegierte Dinge für Unterprivilegierte? Die Kunstkammer im Waisenhaus." In Dolezel, Godel, Peča and Zaunstöck, *Ordnen – Vernetzen – Vermitteln*, 49–72.

Lauzen, Martin Schwartz. *Jews and Christians in Denmark: From the Middle Ages to Recent Times*. Leiden: Brill, 2015.

Leis, Arlene and Kacie L. Wills, eds. *Women and the Art and Science of Collecting in Eighteenth-Century Europe*. New York: Routledge, 2021.

Lenders, Rob. "The Historical Truth behind the 'Salmon-Servant' Myth." In Smith and Egmond, *Ichthyology in Context*, 454–473.

Leong, Elaine. *Recipes and Everyday Knowledge: Medicine, Science, and the Household in Early Modern England*. Chicago: Chicago University Press, 2018.

Lesch, John E. "Systematics and the Geometrical Spirit." In Frängsmyr, Heilbron and Rider, *The Quantifying Spirit*, 73–111.

Lesser, Ludwik. *Chronik der Gesellschaft der Freunde in Berlin*. Berlin: Petsch, 1842.

Lesser, Richard. "Dr. Marcus Elieser Bloch: Ein Jude begründet die moderne Ichthyologie." In *Haskala: Die jüdische Aufklärung in Deutschland 1769–1812*, edited by Christoph Schulte, 238–246. Wolfenbüttel: Wallstein, 1999.

Leu, Urs and Peter Opitz, eds. *Conrad Gessner (1516–1565): Die Renaissance der Wissenschaften/The Renaissance of Learning*. Zurich: De Gruyter Oldenbourg, 2019.

Levitin, Dmitri. *Ancient Wisdom in the Age of the New Science: Histories of Philosophy in England, c. 1640–1700*. Cambridge: Cambridge University Press, 2015.

Lewis, Eric. "Aristotle's Physical Theory." In *The Cambridge History of Science*, vol. 1: *Ancient Science*, edited by Jones Alexander and Liba Taub, 196–214. Cambridge: Cambridge University Press, 2018.

Lewis, Gillian. "The Debt of John Ray and Martin Lister to Guillaume Rondelet of Montpellier." *Notes and Records of the Royal Society of London* 66, no. 4 (2012): 323–339.

Liebau, Heike. *Cultural Encounters in India: The Local Co-Workers of Tranquebar Mission, 18th to 19th Centuries*. London: Routledge, 2017.

Lindroth, Sten. "The Two Faces of Linnaeus." In *Linnaeus, The Man and His Work*, edited by Tore Frängsmyr, 1–62. Berkeley: University of California Press, 1983.

Long, Pamela O. *Artisan/Practitioners and the Rise of the New Sciences, 1400–1600*. Corvallis: Oregon State University Press, 2011.

Lönnberg, Einar. *Peter Artedi: A Bicentenary Memoir Written on Behalf of the Swedish Royal Academy of Science*, trans. W.E. Harlock. Uppsala: Almqvist & Wiksells, 1905.

Lowenstein, Steven M. *The Berlin Jewish Community: Enlightenment, Family, and Crisis, 1770–1830*. Oxford: Oxford University Press, 1994.

Lux, David S. and Harold J. Cook. "Closed Circles or Open Networks? Communicating at a Distance During the Scientific Revolution." *History of Science* 36, no. 2 (1998): 179–211.

MacGregor, Arthur. "Patrons and Collectors: Contributors of Zoological Subjects to the Works of George Edwards (1694–1773)." *Journal for the History of Collections* 25, no. 1 (2013): 35–44.

MacGregor, Arthur, ed. *Naturalists in the Field: Collecting, Recording and Preserving the Natural World from the Fifteenth to the Twenty-First Century*, Leiden: Brill, 2018.

MacGregor, Arthur. "European Enlightenment in India: An Episode of Anglo-German Collaboration in the Natural Sciences on the Coromandel Coast." In MacGregor, *Naturalists in the Field*, 365–392.

Maclean, Ian. "White Crows, Graying Hair, and Eyelashes: Problems for Natural Historians in the Reception of Aristotelian Logic and Biology from Pomponazzi to Bacon." In Pomata and Siraisi, *Historia*, 147–179.

Margócsy, Dániel. "'Refer to folio and number': Encyclopedias, the Exchange of Curiosities, and Practices of Identification before Linnaeus." *Journal of the History of Ideas* 71, no. 1 (2010): 63–89.

Margócsy, Dániel. *Commercial Visions: Science, Trade, and Visual Culture in the Dutch Golden Age*. Chicago: Chicago University Press, 2014.

Mariss, Anne. *A World of New Things: Praktiken der Naturgeschichte bei Johann Reinhold Forster*. Frankfurt am Main: Campus, 2015.

Marr, Alexander. "Knowing Images." *Renaissance Quarterly* 69, no. 3 (2016): 1000–1013.

Marroquín, Jaime Arredondo and Ralph Bauer, eds. *Translating Nature: Cross-Cultural Histories of Early Modern Science*. Philadelphia: University of Pennsylvania Press, 2019.

Meijers, Debora J. "The Paper Museum as a Genre: The Corpus of Drawings in St Petersburg within a European Perspective." In *The Paper Museum of the Academy of Sciences in St Petersburg c. 1725–1760*, edited by Renée Kistemaker, Natalya Kopaneva, Debora J. Meijers and Georgy Vilinbakhof, 19–54. Amsterdam: Royal Netherlands Academy of Arts and Sciences, 2005.

Mendelsohn, J. Andrew. "The World on a Page: Making a General Observation in the Eighteenth Century." In Daston and Lunbeck, *Histories of Scientific Observation*, 69–89.

Merriman, Peter. "Peter Artedi – Systematist and Ichthyologist." *Copeia* 25, no. 1 (1938): 33–39.

Merriman, Peter. "A Rare Manuscript Adding to our Knowledge of the Work of Peter Artedi." *Copeia* 2, no. 1 (1941): 64–69.

Michell, A.R. "The European Fisheries in Early Modern History." In *The Cambridge Economic History of Europe*, edited by E.E. Rich and C.H. Wilson, 133–184. Cambridge: Cambridge University Press, 1977.

Miert, Dirk van. "Introduction." In *Communicating Observations in Early Modern Letters (1500–1675): Epistolography and Epistemology in the Age of the Scientific Revolution*, edited by Dirk van Miert, 1–8. London: The Warburg Institute, 2013.

Mijnhardt, Wijnand. "'The World We Have Lost': In Praise of a Comprehensive Concept of Science and Scholarship." In Bergvelt and Meijers, *Teyler's Foundation in Haarlem*, 69–86.

Miller, Lulu. *Why Fish Don't Exist: A Story of Loss, Love and the Hidden Order of Life*. New York: Simon & Schuster, 2020.

Miracle, Maria Eulàlia Gassó. *Coenraad Jacob Temminck and the Emergence of Systematics (1800–1850)*. Leiden: Brill, 2022.

Monteiro, Carolina, and Erik Odegard. "Slavery at the Court of the 'Humanist Prince' Reexamining Johan Maurits van Nassau-Siegen and his Role in Slavery, Slave Trade and Slave-Smuggling in Dutch Brazil." *Journal of Early American History* 10, no. 1 (2020): 3–32.

Montgomerie, Robert and Tim R. Birkhead. "Samuel Pepys's Hand-Coloured Copy of John Ray's 'The Ornithology of Francis Willughby' (1678)." *Journal of Ornithology* 150, no. 4 (2009): 883–891.

Moore, Alec B.M. and Jan G. Hiddink. "Identifying Critical Habitat With Archives: 275-Year-Old Naturalist's Notes Provide High-Resolution Spatial Evidence of Long-Term Core Habitat for A Critically Endangered Shark." *Biological Conservation* 272 (2022).

Moxham, Noah. "Edward Tyson's *Phocaena*: A Case Study in the Institutional Context of Scientific Publishing." *Notes and Records of the Royal Society in London*, 66, no. 3 (2012): 235–252.

Moxham, Noah. "Authors, Editors and Newsmongers: Form and Genre in the Philosophical Transactions under Henry Oldenburg." In Raymond and Moxham, *News Networks in Early Modern Europe*, 465–492.

Mulholland, Richard. "The Mechanism and Materials of Painting Colour *ad vivum* in the Eighteenth Century." In Balfe, Woodall and Zittel, *Ad vivum?*, 328–355.

Müller, Johannes. "Distance, Geography, and Anecdote in M.E. Bloch's Natural History of Fishes." In Smith and Egmond, *Ichthyology in Context*, 612–631.

Müller-Wille, Staffan. "Collection and Collation: Theory and Practice of Linnaean Botany." *Studies in History and Philosophy of the Biological and Biomedical Sciences* 38, no. 3 (2007): 541–562.

Müller-Wille, Staffan. "Collection and Collation: Theory and Practice of Linnaean Botany." *Studies in History and Philosophy of the Biological and Biomedical Sciences* 38, no. 3 (2007): 541–562.

Müller-Wille, Staffan. "Linnaean Paper Tools." In Curry, Jardine, Secord, and Spary, *Worlds of Natural History*, 205–220.

Müller-Wille, Staffan. "Names and Numbers: 'Data' in Classical Natural History, 1758–1859." *Osiris* 32 (2017): 109–208.

Müller-Wille, Staffan. "Walnuts at Hudson Bay, Coral Reefs in Gotland: The Colonialism of Linnean Botany." In *Colonial Botany: Science, Commerce, and Politics in the Early Modern World*, edited by Londa Schiebinger and Claudia Swan 34–48. Philadelphia: University of Pennsylvania Press, 2005.

Müller-Wille, Staffan and Karen Reeds. "A Translation of Carl Linnaeus's Introduction to *Genera plantarum* (1737)." *Studies in History and Philosophy of Biological and Biomedical Sciences* 38, no. 3 (2007): 563–572.

Müller-Wille, Staffan and Sara T. Scharf. "Indexing Nature: Carl Linnaeus (1707–1778) and His Fact-Gathering Strategies." *Svenska Linnesällskapets Årsskrift* 94 (2011): 31–60.

Munck, Thomas. "Eighteenth-Century Review Journals and the Internationalization of the European Book Market." *The International History Review* 32, no. 3 (2010): 415–435.

Near, Thomas J. and Christine E. Thacker. "Phylogenetic Classification of Living and Fossil Ray-Finned Fishes (Actinopterygii)." *Bulletin of the Peabody Museum of Natural History* 65, no. 1 (2024): 340–357.

Neri, Janice. *The Insect and the Image: Visualizing Nature in Early Modern Europe 1500–1700*. Minneapolis: University of Minnesota Press, 2011.

Nickelsen, Kärin. *Draughtsmen, Botanists and Nature: The Construction of Eighteenth-Century Botanical Illustrations*. Dordrecht: Springer, 2006.

Nissen, Claus. *Die Zoologische Buch-Illustration*, vol. 2. Stuttgart: Anton Hiersemann, 1978.

Novikoff, Alex J. "Toward a Cultural History of Scholastic Disputation." *American Historical Review* 117, no. 2 (2012): 331–364.

Nummedal, Tara and Paula Findlen. "Words of Nature: Scientific Books in the Seventeenth Century." In *Thorton and Tully's Scientific Books, Libraries and Collectors: A Study of Bibliography and the Book Trade in Relation to the History of Science*, edited by Andrew Hunter, 164–215. Aldershot: Ashgate, 2000.

Nybelin, Orvar. "Tvenne Opublicerade Artedi-Manuskript." *Svenska Linnésällskapets Årsskrift* 18 (1935): 58–77.

Nybelin, Orvar. "Kring Petrus Artedi's vistelse i England 1734–1735," *Svenska Linnésällskapets Årsskrift* 49 (1966): 9–27.

Ochs, Kathleen H. "The Royal Society of London's History of Trades Programme: An Early Episode in Applied Science." *Notes and Records of the Royal Society* 39, no. 2 (1985): 129–158.

Ogilvie, Brian. *The Science of Describing: Natural History in Renaissance Europe*. Chicago: University of Chicago Press, 2006.

Ogilvie, Brian. "Natural History, Ethics, and Physico-Theology." In Pomata and Siraisi, *Historia*, 75–103.

Ogilvie, Brian. "Insects in John Ray's Natural History and Natural Theology." In Enenkel and Smith, *Zoology in Early Modern Culture*, 234–260.

Ogilvie, Brian. "Willughby on Insects." In Birkhead, *Virtuoso by Nature*, 335–359.

Ogilvie, Brian. "Visions of Ancient Natural History." In Curry, Jardine, Secord, and Spary, *Worlds of Natural History*, 17–32.

Olesko, Kathryn. "Germany." In *The Cambridge History of Science: Modern Science in National, Transnational, and Global Context*, edited by Hugh Richard Slotten, Ronald L. Numbers and David N. Livingstone, 233–277. Cambridge: Cambridge University Press, 2020.

Omodeo, Pietro Daniel. "The Invisible Fisherman: The Economy of Water Knowledge in Early Modern Venice." In Smith and Egmond, *Ichthyology in Context*, 364–391.

Ong, Walter. *Ramus, Method and the Decay of the Dialogue: From the Art of Discourse to the Art of Reason*. Cambridge, Mass.: Harvard University Press, 1958.

Overduin-de Vries, Anne M. and Paul J. Smith. "Fishing in the Past: Biodiversity, Art History, and Citizen Science – Preliminary Results." In Smith and Egmond, *Ichthyology in Context*, 298–321.

Paepke, Hans-Joachim. *Bloch's Fish Collection in the Museum für Naturkunde der Humboldt-Universität zu Berlin: An Illustrated Catalog and Historical Account*. Ruggell: A.R. Gantner, 1999.

Paepke, Hans-Joachim. "Ein jüdischer Untertan des Preußenkönings Friedrich II. studiert die Fischfauna der Welt." In *Klasse, Ordnung, Art: 200 Jahre Museum für Naturkunde*, edited by Ferdinand Damaschun, Sabine Hackethal, Hannelore Landsberg, and Reinhold Leinfelder, 84–87. Rangsdorf: Basiliskenpresse, 2010.

Palm, Lodewijk. "Leeuwenhoek and Other Dutch Correspondents of the Royal Society." *Notes and Records of the Royal Society of London* 43, no. 2 (1989): 192–193.

Park, Katharine and Lorraine Daston. "Introduction: The Age of the New." In *The Cambridge History of Science*, vol. 3: *Early Modern Science*, edited by Katharine Park and Lorraine Daston, 1–17. Cambridge: Cambridge University Press, 2006.

Parsons, Christopher M., and Kathleen S. Murphy. "Ecosystems under Sail: Specimen Transport in the Eighteenth-Century French and British Atlantics." *Early American Studies* 10, no. 3 (2012): 503–529.

Pastorino, Cesare. "Weighing Experience: Experimental Histories and Francis Bacon's Quantitative Program." *Early Science and Medicine* 16, no. 6 (2011): 542–570.

Pauly, Daniel. "Anecdotes and the Shifting Baseline Syndrome of Fisheries." *Trends in Ecology and Evolution* 10, no. 10 (1995): 430.

Pedersen, Tara E. *Mermaids and the Production of Knowledge in Early Modern England.* Farnham: Ashgate, 2015.

Péquignot, Amandine. "Une peau entre deux feuilles, l'usage de l'herbier en taxidermie aux XVIIIe et XIXe siècles en France." *Revue d'histoire des sciences* 59, no. 1 (2006): 127–136.

Petukhova, Tatyana. "A History of Fish Glue as an Artist's Material: Applications in Paper and Parchment Artifacts." *The Book and Paper Group Annual* 19 (2000): 111–114.

Phillips, Denise. *Acolytes of Nature: Defining Natural Science in Germany, 1770–1850.* Chicago: Chicago University Press, 2012.

Pickering, Victoria. "Mobilising Historical Botanical Data as Research." *Nuncius* 39, no. 3 (2024): 759–774.

Pietsch, Theodore W. *The Curious Death of Peter Artedi: A Mystery in the History of Science*. New York: Scott & Nix Inc., 2010.

Pietsch, Theodore W. *Charles Plumier (1646–1704) and His Drawings of French and Caribbean Fishes*. Paris: Publications scientifiques du Muséum, 2017.

Pietsch, Theodore W. and Hans Aili. "Peter Artedi's Catalogue of the Fishes of the Baltic Sea: An English Translation with an Introduction and Commentary," *Zoological Journal of the Linnean Society* 189, no. 3 (2020): 975–997.

Pietsch, Theodore W. and Hans Aili. "Swedish Naturalist Peter Artedi (1705–1735) and His Place in the History of Biosystematics as Exemplified by his *Ichthyologia sive opera omnia piscibus* of 1738." *Zootaxa* 5169, no. 6 (2022): 589–598.

Pietsch, Theodore W. and Hans Aili. "Peter Artedi's 'Manuscriptum ichthyologicum', A Source for Albertus Seba's *Locupletissimi rerum naturalium thesauri accurata descriptio* (1759)." *Archives of Natural History* 50, no. 1 (2023): 118–132.

Pietsch, Theodore W. and Hans Aili. "Peter Artedi's Early Observations of the Spotted Hyena and Other Exotic Animals During a Visit to London (1734–1735)." *Archives of Natural History* 50, no. 2 (2023): 410–416.

Pietsch, Theodore W. and Hans Aili. "Peter Artedi's 'Idea institutionum Trichozoologiae' and the Classification of Mammals." *Journal of Natural History* 57, no. 17–20 (2023): 1066–1079.

Pietsch, Theodore W. and Justin R. Hanisch. "Louis Renard (1678/1679–1746) and His *Poissons, ecrevisses et crabes* (1719): 300 Years of One of Natural History's Most Curious Colour Plate Books." In Smith and Egmond, *Ichthyology in Context*, 584–611.

Podda, Cinzia, Francesco Palmas, Antonio Pusceddu and Andrea Sabatini, "When the Eel Meets Dams: Larger Dams' Long-Term Impacts on *Anguilla anguilla* (L., 1758)." *Frontiers in Environmental Science* 10 (2022): 1–10.

Pomata, Gianna. "Observation Rising: Birth of an Epistemic Genre, ca. 1500–1650." In *Histories of Scientific Observation*, edited by Lorraine Daston and Elizabeth Lunbeck, 45–80. Chicago: University of Chicago Press, 2011.

Pomata, Gianna. "*Praxis Historialis*: The Uses of Historia in Early Modern Medicine." In Pomata and Siraisi, *Historia*, 105–146.

Pomata, Gianni and Nancy G. Siraisi. "Introduction." In *Historia: Empiricism and Erudition in Early Modern Europe*, edited by Gianna Pomata and Nancy G. Siraisi, 1–38. Cambridge, Mass.: MIT Press, 2005.

Poole, William. "The Willughby Library in the Time of Francis the Naturalist." In Birkhead, *Virtuoso by Nature*, 227–243.

Porter, Roy. *The Making of Geology: Earth Science in Britain, 1660–1815*. Cambridge: Cambridge University Press, 1977.

Porter, Roy. "Introduction." In *The Cambridge History of Science*, vol. 4, *Eighteenth-century Science*, edited by Roy Porter, 1–20. Cambridge: Cambridge University Press, 2003.

Prieto, Andrés I. *Missionary Scientists: Jesuit Science in Spanish South America, 1570–1810*. Nashville: Vanderbilt University Press, 2011.

Pumfrey, Steven. "Ideas above His Station: A Social Study of Hooke's Curatorship of Experiments." *History of Science* 29, no. 1 (1991): 1–44.

Pumfrey, Steven. "Who did the Work? Experimental Philosophers and Public Demonstrators in Augustan England." *British Journal for the History of Science* 28, no. 2 (1995): 131–156.

Raj, Kapil. *Relocating Modern Science: Circulation and the Construction of Scientific Knowledge in South Asia and Europe*. Basingstoke: Palgrave Macmillan, 2007.

"Ralph Johnson, Vicar of Brignall 1656–1695." In *Teesdale Record Society* 15 (1945): 9–32.

Raymond, Joad and Noah Moxham, eds. *News Networks in Early Modern Europe*. Leiden: Brill, 2016.

Raven, Charles E. *John Ray, Naturalist: His Life and Works*. Cambridge University Press, Cambridge, 1942.

Rees, Joachim. *Die Verzeichnete Fremde. Formen und Funktionen des Zeichnens im Kontext europäischer Forschungsreisen 1770–1830*. Paderborn: Wilhelm Fink, 2015.

Remmert, Volker R. "'Docet parva picture, quod multae scripturae non dicunt.' Frontispieces, their functions, and their audiences in seventeenth-century mathematical sciences," in *Transmitting Knowledge: Words, Images, and Instruments in Early Modern Europe*, edited by Sachiko Kusukawa and Ian Maclean, 239–270. Oxford University Press, Oxford, 2006.

Remmert, Volker R. *Picturing the Scientific Revolution*. Philadelphia: Saint Joseph's University Press, 2011.

Renn, Jürgen and Florian Schmaltz. "Institutions and Knowledge Systems: Theoretical Perspectives." In *The Institutionalization of Science in Early Modern Europe*, edited by Mordechai Feingold and Giulia Giannini, 292–296. Leiden: Brill, 2019.

Rijks, Marlise. "'Unusual Excrescences of Nature': Collected Coral and the Study of Petrified Luxury in Early Modern Antwerp." *Dutch Crossing: Journal of Low Countries Studies* 43, no. 2 (2019): 54–83.

Rijks, Marlise. "Scales, Skins, and Carapaces in Antwerp Collections." In *The Matter of Mimesis: Studies on Mimesis and Materials in Nature, Art and Science*, edited by Marjolein Bol and Emma C. Spary, 295–320. Leiden: Brill, 2023.

Rijks, Marlise. "A Taste for Fish: Paintings of Aquatic Animals in the Low Countries (1560–1729)." In Smith and Egmond, *Ichthyology in Context*, 259–297.

Ritvo, Harriet. "New Presbyter or Old Priest? Reconsidering Zoological Taxonomy in Britain, 1750–1840," *History of the Human Sciences* 3, no. 2 (1990): 259–276.

Roberts, Lissa. "Introduction: Centres and Cycles of Accumulation." In *Centres and Cycles of Accumulation in and Around the Netherlands During the Early Modern Period*, edited by Lissa Roberts, 3–27. Berlin: LIT, 2011.

Robinson, Henry W. and Walter Adams, eds. *The Diary of Robert Hooke, 1672–1680*. London: Taylor and Francis, 1935.

Robles, Whitney Barlow. "Natural History in Two Dimensions." *Common Place: The Journal of Early American Life* 18, no. 1 (2018), online publication https://commonplace.online/article/vol-18-no-1-robles/.

Robles, Whitney Barlow. "Flatness." In *The Philosophy Chamber: Art and Science in Harvard's Teaching Cabinet, 1766–1820*, edited by Ethan W. Lasser, 191–209. New Haven: Yale University Press, 2017.

Robles, Whitney Barlow. *Curious Species: How Animals Made Natural History*. New Haven: Yale University Press, 2023.

Roling, Bernd. *Drachen und Sirenen: Die Aufarbeitung und Abwicklung der Mythologie an den europäischen Universitäten*. Leiden: Brill, 2010.

Roos, Anna Marie. *Web of Nature: Martin Lister (1639–1712), the First Arachnologist*. Leiden: Brill, 2011.

Roos, Anna Marie. *Martin Lister and His Remarkable Daughters: The Art of Science in the Seventeenth Century*. Oxford: The Bodleian Library, 2019.

Rose, Edwin. "Specimens, Slips and Systems: Daniel Solander and the Classification of Nature at the World's First Public Museum, 1753–1768." *British Journal for the History of Science* 51, no. 2 (2018): 205–237.

Rousseau, Georg Sebastian and Roy Porter, eds. *The Ferment of Knowledge: Studies in the Historiography of Eighteenth-Century Science*. Cambridge: Cambridge University Press, 1980.

Rudwick, Martin. *Rudwick, Bursting the Limits of Time: The Reconstruction of Geohistory in the Age of Revolution*. Chicago: Chicago University Press, 2005.

Rudwick, Martin. "Georges Cuvier's Paper Museum of Fossil Bones." *Archives of Natural History* 27, no. 1 (2010): 51–68.

Ruhland, Thomas. *Pietistische Konkurrenz und Naturgeschichte: Die Südasienmission der Herrnhuter Brüdergemine und die Dänisch-English-Hallesche Mission (1755–1802)*. Herrnhut: Herrnhuter Verlag, 2018.

Saénz–Arroyo, Andrea, Callum M. Roberts, Jorge Torre and Micheline Carinõ-Olvera. "Using Fishers' Anecdotes, Naturalists' Observations and Grey Literature to Reassess Marine Species at Risk: The Case of the Gulf Grouper in the Gulf of California, Mexico." *Fish and Fisheries* 6, no. 2 (2005): 121–133.

Savage, Elizabeth and Ad Stijnman. "'Material Colour': The Heritage of Colour Knowledge in Seventeenth- and Eighteenth-Century Printshops." In *Colour Histories: Science, Art, and Technology in the 17th and 18th Centuries*, edited by Magdalena Bushart and Friedrich Steinle, 95–113. Berlin: De Gruyter, 2015.

Scharf, Sara T. "Identification Keys, the 'Natural Method,' and the Development of Plant Identification Manuals." *Journal of the History of Biology* 42, no. 1 (2009): 73–117.

Scearce, Caroline. "Fish Facts: Disciplinary Development of Ichthyology in Nineteenth-Century Europe." PhD diss., University of Oklahoma, 2019.

Schiebinger, Londa. "Why Mammals are Called Mammals: Gender Politics in Eighteenth-Century Natural History." *The American Historical Review* 98, no. 2 (1993): 382–411.

Schiebinger, Londa. "Women of Natural Knowledge." In Park and Daston, *Cambridge History of Science*, vol. 3, 192–205.

Schmidt, Benjamin. *Inventing Exoticism: Geography, Globalism, and Europe's Early Modern World*. Philadelphia: University of Pennsylvania Press, 2015.

Secord, Anne. "Coming to Attention: A Commonwealth of Observers during the Napoleonic Wars." In Daston and Lunbeck, *Histories of Scientific Observation*, 421–444.

Secord, James. "Knowledge in Transit." *Isis* 95, no. 4 (2004): 654–672.

Seehafer, Michèle. "Shimmering Virtue: Joris Hoefnagel and the Uses of Shell Gold in the Early Modern Period." In *Materialized Identities: Objects, Affects, and Effects in Early Modern Culture, 1450–1750*, edited by Susanna Burghartz, Lucas Burkart,

Christine Göttler and Ulinka Rublack, 281–321. Amsterdam: Amsterdam University Press, 2021.

Serjeantson, Richard. "The Education of Francis Willughby." In Birkhead, *Virtuoso by Nature*, 44–98.

Shapin, Steven. "The Invisible Technician." *American Scientist* 77, no. 6 (1989): 544–563.

Shapin, Steven. *A Social History of Truth: Civility and Science in Seventeenth-Century England*. Chicago: Chicago University Press, 1994.

Shapin, Steven. *Never Pure: Historical Studies of Science as if it Was Produced by People with Bodies, Situated in Time, Space, Culture, and Society, and Struggling for Credibility and Authority*. Baltimore: Johns Hopkins University Press, 2010.

Shapin, Steven and Simon Schaffer. *Leviathan and the Air-Pump: Hobbes, Boyle and the Experimental Life*. Princeton: Princeton University Press, 1985.

Shapiro, Barbara. *A Culture of Fact: England 1550–1720*. Ithaca: Cornell University Press, 2000.

Slaughter, Mary. *Universal Language and Scientific Taxonomy in the Seventeenth Century*. Cambridge: Cambridge University Press, 1982.

Sliggers, Bert. "De Verzamelwoede van Martinus van Marum (1750–1837) en de Ouderdom van de Aarde: Herkomst en Functie van het Paleontologisch en Mineralogisch Kabinet van het Teylers Museum." PhD diss., Leiden University, 2017.

Sloan, Kim. *A New World: England's First View of America*. Chapel Hill: University of North Carolina Press, 2007.

Smith, Pamela. *The Body of the Artisan: Art and Experience in the Scientific Revolution*. Chicago: University of Chicago Press, 2004.

Smith, Pamela, ed. *The Varied Role of the Amateur in Early Modern Europe*, special issue of *Nuncius* 31, no. 3 (2016): 485–609.

Smith, Paul J. "Marcgraf's Fish in the Historia Naturalis Brasiliae and the Rhetorics of Autoptic Testimony." In *Toward an Intercultural Natural History of Brazil. The* Historia Naturalis Brasiliae *Reconsidered*, edited by Mariana C. Françozo, 122–141. New York: Routledge, 2023.

Smith, Paul J. and Didi van Trijp. "Dynamiques européennes de l'humanisme érudit dans l'histoire naturelle. Le cas de l'ichtyologie, de Belon, Rondelet et Gessner à Willughby et Ray." In *L'humanisme à l'épreuve de l'Europe (XVe–XVIe siècles)*, edited by Denis Crouzet, Elisabeth Crouzet-Pavan, Philippe Desan and Clémence Revest, 167–181. Ceyzérieu: Champ Vallon, 2019.

Smith, Paul J. and Florike Egmond, eds. *Ichthyology in Context (1500–1880)*. Leiden: Brill, 2023.

Spary, Emma. "Of Nutmegs and Botanists: The Colonial Cultivation of Botanical Identity." In Schiebinger and Swan, *Colonial Botany*, 187–203.

Spary, Emma. "The Naturalist Collecting Community in Paris, 1760–1789: A Preliminary Survey." In Dolezel, Godel, Peča and Zaunstöck, *Ordnen – Vernetzen – Vermitteln*, 297–324.

Stearn, William T. "Carl Linnaeus and the Theory and Practice of Horticulture." *Taxon* 25, no. 1 (1976): 21–31.

Stewart, Larry. "Global Pillage: Science, Commerce, and Empire." In Porter, *Cambridge History of Science*, vol. 4, 825–844.

Stichweh, Rudolf. "The Sociology of Scientific Disciplines: On the Genesis and Stability of the Disciplinary Structure of Modern Science." *Science in Context* 5, no. 1 (1992): 3–15.

Striekwold, Robbert. "Images, Specimens, and Species: Hermann Schlegel on the Various Ways of Depicting a Fish." In Smith and Egmond, *Ichthyology in Context*, 701–726.

Swann, Marjorie. *Curiosities and Texts: The Culture of Collecting in Early Modern England*. University of Pennsylvania Press: Philadelphia, 2001.

Thode Jensen, Niklas. "The Tranquebarian Society: Science, Enlightenment and Useful Knowledge in the Danish-Norwegian East Indies, 1768–1813." *Scandinavian Journal of History* 40, no. 4 (2015): 535–561.

Trepp, Anne-Charlott. "Matters of Belief and Belief that Matters: German Physico-Theology, Protestantism, and the Materialized Word of God in Nature." In Blair and Von Greyerz, *Physico-Theology*, 127–140.

Trijp, Didi van. "Fresh Fish: Observation up Close in Late Seventeenth-Century England." *Notes and Records of the Royal Society of London* 75, no. 3 (2021): 311–332.

Trijp, Didi van and Robbert Striekwold. "The Ichthyologist's Garden." The Recipes Project (2017), https://recipes.hypotheses.org/9798.

Van Dyke, Fred. *Conservation Biology: Foundations, Concepts, Applications*. Dordrecht: Springer, 2008.

Vanzo, Alberto, ed. *Experience in Natural Philosophy and Medicine*, special issue of *Perspectives on Science* 24, no. 3 (2016) 255–379.

Vermij, Rienk. "On the Function of Dedications in Early Modern Scientific Books." *Nuncius* 33, no. 2 (2018): 171–197.

Vliet, Rietje van. *Elie Luzac (1721–1796): Boekverkoper van de Verlichting*. Nijmegen: Vantilt, 2005.

Weber, Andreas and Esther Turnhout. "A Langur from Sumatra: Digital Futures, Material Presents and Colonial Pasts." *Nuncius* 39, no. 3 (2024): 776–788.

Weeks, Sophie. "The Role of Mechanics in Bacon's Great Instauration." In *Philosophies of Technology: Francis Bacon and his Contemporaries*, edited by Claus Zittel, Gisela Engel, Romano Nanni and Nicole C. Karafyllis, 133–195. Leiden: Brill, 2008.

Wells, Ellen B. "M.E. Bloch's *Allgemeine Naturgeschichte der Fische*: A Study." *Archives of Natural History* 10, no. 1 (1981): 7–13.

Wheeler, Alwhyne. "Petrus Artedi, Founder of Modern Ichthyology." In Kullander and Fernholm, *Proceedings of the Fifth Congress on European Ichthyology*, 3–10.

Wheeler, Alwhyne. "The Linnaean Fish Collection in the Linnean Society of London." *Zoological Journal of the Linnean Society* 84, no. 1 (1985): 1–76.

Wheeler, Alwhyne. "The Linnaean Fish Collection in the Zoological Museum of the University of Uppsala." *Zoological Journal of the Linnean Society* 103, no. 2 (1991): 145–195.

Whitehead, Peter. "George Markgraf and Brazilian Zoology." In *Johan Maurits van Nassau-Siegen: A Humanist Prince in Europe and Brazil*, edited by E. van den Boogaard, 424–471. The Hague: Mauritshuis, 1979.

Whitehead, Peter and Marinus Boeseman, eds. *A Portrait of Dutch 17th Century Brasil: Animals, Plants and People by the Artists of Johan Maurits of Nassau*. Amsterdam: North-Holland Publishing Company, 1989.

Winsor, Mary P. "Linnaeus's Biology was not Essentialist." *Annals of the Missouri Botanical Garden* 93, no. 1 (2006): 2–7.

Winterbottom, Anna. *Hybrid Knowledge in the Early East India Company World*. Basingstoke: Palgrave Macmillan, 2015.

Wragge-Morley, Alexander. *Aesthetic Science: Representing Nature in the Royal Society of London, 1650–1720*. Chicago: Chicago University Press, 2020.

Yale, Elizabeth. *Sociable Knowledge: Natural History and the Nation in Early Modern Britain*. Philadelphia: University of Pennsylvania Press, 2016.

Yeo, Richard. "Classifying the Sciences." In Porter, *Cambridge History of Science*, vol. 4, 241–266.

Zaunstöck, Holger, ed. *Gebaute Utopien, Franckes Schulstadt in der Geschichte europaischer Stadtentwürfe*. Halle: Frankesche Stiftungen, 2010.

Zimmer, Katarina. "What's in a Fish's Name?," *Nautilus* (12 March 2024), https://nautil.us/whats-in-a-fishs-name-530331/?trk=feed_main-feed-card_feed-article-content.

Index

Note: this index contains scientific names for species in both Latin and English, e.g. "*Anguilla anguilla*" and "angel shark"; it also contains non-scientific names in English (such as cod), Swedish (for example *Slom*) and Tamil (e.g. *Karamudel*). Any non-English names have been italicised.

www.ingramcontent.com/pod-product-compliance
Lightning Source LLC
Chambersburg PA
CBHW071333230926
51849CB00002B/385
* 9 7 8 9 0 0 4 5 4 9 4 9 4 *